EAA series - Textbook

EAA series is successor of the EAA Lecture Notes and supported by the European Actuarial Academy (EAA GmbH), founded on the 29 August, 2005 in Cologne (Germany) by the Actuarial Associations of Austria, Germany, the Netherlands and Switzerland. EAA offers actuarial education including examination, permanent education for certified actuaries and consulting on actuarial education.

actuarial-academy.com

EAA series/EAA Lecture Notes

Wüthrich, M.V.; Bühlmann, H.; Furrer, H. **Market-Consistent Actuarial Valuation** 2007

Lütkebohmert, E. **Concentration Risk in Credit Portfolios** 2009

Sundt, B.; Vernic, R. **Recursions for Convolutions and Compound Distributions with Insurance Applications** 2009

Ohlsson, E.; Johansson, B. **Non-Life Insurance Pricing with Generalized Linear Models** 2010

Esbjörn Ohlsson · Björn Johansson

Non-Life Insurance Pricing with Generalized Linear Models

Dr. Esbjörn Ohlsson
Länsförsäkringar Alliance
106 50 Stockholm
Sweden
ESBJ@math.su.se

Dr. Björn Johansson
Länsförsäkringar Alliance
106 50 Stockholm
Sweden

ISSN 1869-6929
e-ISSN 1869-6937
ISBN 978-3-642-10790-0
e-ISBN 978-3-642-10791-7
DOI 10.1007/978-3-642-10791-7
Springer Heidelberg Dordrecht London New York

Library of Congress Control Number: 2010923536

Mathematics Subject Classification (2000): 91Gxx, 62Jxx, 97M30

Cover design: WMXDesign GmbH, Heidelberg

Printed on acid-free paper

Springer is part of Springer Science+Business Media (www.springer.com)

To our families

Preface

Non-life insurance pricing is the art of setting the price of an insurance policy, taking into consideration various properties of the insured object and the policy holder. The main source on which to base the decision is the insurance company's own historical data on policies and claims, sometimes supplemented with data from external sources. In a *tariff analysis*, the actuary uses this data to find a model which describes how the claim cost of an insurance policy depends on a number of explanatory variables. In the 1990's British actuaries introduced generalized linear models (GLMs) as a tool for tariff analysis and this has now become the standard approach in many countries.

This book focuses on methods based on GLMs that we have found useful in our actuarial practice, and intend to provide a set of tools that fills most needs for a tariff analysis. Successive versions of the text have been the basis for a course in non-life insurance mathematics at Stockholm University since 2001. This course is part of the curriculum set up by the *Swedish Actuarial Society* to meet the European *Core Syllabus* for actuarial education.

The aim is to present the basic theory of GLMs in a tariff analysis setting, and also to give some useful extensions of standard GLM theory that are not in common use, viz. the incorporation of random effects and the use of smoothing splines. Random effect models can be used to great advantage for handling categorical variables with a large number of possible levels, and there is an interesting connection with the traditional actuarial field of credibility theory that will be developed here. Smoothing splines is a powerful method for modeling the effect of continuous variables; such analyses are often presented under the name generalized additive models (GAMs). While GAMs have been used in biostatistics for several years, they have not yet found their way into the actuarial domain to any large extent.

The text is intended for practicing actuaries and actuarial students with a background in mathematics and mathematical statistics: proofs are included whenever possible without going into asymptotic theory. The prerequisites are basic university mathematics—including a good knowledge of linear algebra and calculus—and basic knowledge of probability theory, likelihood-based statistical inference and regression analysis. A second course in probability, say at the level of Gut [Gu95], is useful.

In order to provide students with the possibility to work with real data of some complexity, we have compiled a collection of data sets to be used in connection with a number of case studies. The data was provided by Länsförsäkringar Alliance and is available at www.math.su.se/GLMbook.

For working through the case studies, a suitable software is needed. In Appendix we have given some hints on how to proceed using the SAS system, since this is a standard software at many insurance companies. Matlab and R are other possibilities for a course based on this text. There are also some good specialized software packages for tariff analysis with GLMs on the market, but these are not generally available to students; we also believe that having to write your own programs helps in understanding the subject.

Some sections that are not necessary for understanding the rest of the text have been indicated by a star.

We wish to acknowledge the support from Stockholm University, Division of Mathematical Statistics, where we both have been senior lecturers for several years, and the Non-life actuarial group at Länsförsakringar Alliance, were we are now working as actuaries. Special thanks to Professor Rolf Sundberg for providing the justification for the pure premium confidence intervals. Finally, we wish to thank Viktor Grgić for valuable comments and help with the graphics.

Stockholm

Esbjörn Ohlsson
Björn Johansson

Contents

Acronyms

ANOVA	Analysis of variance
BLP	Best linear predictor
CGF	Cumulant-generating function
EDM	Exponential dispersion model
GAM	Generalized additive model
GCV	Generalized cross-validation
GLM	Generalized linear model
HGLM	Hierarchical generalized linear model
LRT	Likelihood ratio test
MLE	Maximum likelihood estimator
MLF	Multi-level factor
MMT	Method of marginal totals
MSEP	Mean square error of prediction
ODP	Overdispersed Poisson distribution
TPL	Third party liability

Chapter 1
Non-Life Insurance Pricing

In this introductory chapter we describe the problem of pricing in non-life insurance, and define some basic concepts and assumptions. We also introduce an example from moped insurance that will be used repeatedly.

A non-life insurance policy is an agreement between an insurance company and a customer—the policyholder—in which the insurer undertakes to compensate the customer for certain unpredictable losses during a time period, usually one year, against a fee, the *premium*. A non-life insurance policy may cover damages on a car, house or other property, or losses due to bodily injury to the policyholder or another person (third party liability); for a company, the insurance may cover property damages, cost for business interruption or health problems for the employees, and more. In effect, any insurance that is not life insurance is classified as *non-life insurance*, also called *general insurance* or (in the US) *property and casualty* insurance. In Germany, the term is quite different: *Schadenversicherung*—a similar name applies in Sweden.

By the insurance contract, economic risk is transferred from the policyholder to the insurer. Due to the law of large numbers, the loss of the insurance company, being the sum of a large number of comparatively small independent losses, is much more predictable than that of an individual (in relative terms): the loss should not be too far from its expected value. This leads us to the generally applied principle that the premium should be based on the expected (average) loss that is transferred from the policyholder to the insurer. There must also be a loading for administration costs, cost of capital, etc., but that is not the subject here.

The need for statistical methods comes from the fact that the expected losses vary between policies: the accident rate is not the same for all policyholders and once a claim has occurred, the expected damages vary between policyholders. Most people would agree that the fire insurance premium for a large villa should be greater than for a small cottage; that a driver who is more accident-prone should pay more for a car insurance; or that installing a fire alarm which reduces the claim frequency should give a discount on the premium. Until the mid 90's, non-life insurance pricing was heavily regulated in Sweden, as in most other western countries. The "fairness" of premiums was supervised by the national FSA (Financial Services Author-

E. Ohlsson, B. Johansson, *Non-Life Insurance Pricing with Generalized Linear Models*, EAA Lecture Notes,
DOI 10.1007/978-3-642-10791-7_1, © Springer-Verlag Berlin Heidelberg 2010

ity), leading to a conform premium structure for all insurance companies on the market.

Today, the market has been deregulated in many countries: the legislation has been modified to ensure free competition rather than uniform pricing. The idea is that if an insurance company charges too high a premium for some policies, these will be lost to a competitor with a more fair premium. Suppose that a company charges too little for young drivers and too much for old drivers; then they will tend to loose old drivers to competitors while attracting young drivers; this *adverse selection* will result in economic loss both ways: by loosing profitable and gaining underpriced policies. The conclusion is that on a competitive market it is advantageous to charge a fair premium, if by fairness we mean that each policyholder, as far as possible, pays a premium that corresponds to the expected losses transferred to the insurance company.

Most often a distinction is made between the overall premium level of an insurance portfolio, and the question of how the total required premium should be divided between the policyholders. The overall premium level is based on considerations on predicted future costs for claims and administration, the capital income on provisions for future costs (reserves) and the cost of capital (expected profit), as well as the market situation. Historically, these calculations have not involved much statistical analysis and they will not be discussed in this text.

On the other hand, the question of how much to charge each individual given the overall premium level involves the application of rather advanced statistical models. Before considering these, we will introduce some fundamental concepts.

1.1 Rating Factors and Key Ratios

For each policy, the premium is determined by the values of a number of variables, the *rating factors*; to estimate this relationship, a statistical model is employed. The rating factors usually belong to one of the following categories:

- *Properties of the policyholders*: age or gender if the policyholder is a private person, line of business for a company, etc.;
- *Properties of the insured objects*: age or model of a car; type of building, etc.;
- *Properties of the geographic region*: per capita income or population density of the policyholder's residential area, etc.

The collection of variables that may be considered is determined by the availability of data. While the age and gender of a policyholder are usually readily accessible, it is much harder to get reliable data on a persons driving behavior or the number of in-bed smokers in a house. There are also limitations in that rating factors must not be considered offensive by the policyholders.

A statistician making contact with non-life insurance pricing for the first time might consider using linear regression models of the relationship between the claim cost and the variables at hand. Traditionally, however, rating factors are formed by dividing variables into classes, e.g. age into age intervals, if they are not already categorical. The reason for this is partly that we seldom have a linear relation and partly the rule by which the premium is computed for a policyholder, the *tariff*. In the old days the tariff was a price list on paper, but today the computations are of course made with a computer, and the tariff is a set of tables by which the premium can be computed for any policy.

In Chap. 5 we will consider methods for using continuous variables directly, but until then we will assume that they are divided into classes before the analysis begins. Policies that belong to the same class for each rating factor are said to belong to the same *tariff cell* and are given the same premium.

Example 1.1 (Moped insurance) This example, which will be used repeatedly in the sequel, is based on data from the Swedish insurance company *Wasa*, before its 1999 fusion with *Länsförsäkringar Alliance*. In Sweden, moped insurance involves three types of cover: TPL (third party liability), partial casco and hull. TPL covers any bodily injuries plus property damages caused to others in a traffic accident. *Partial casco* covers theft but also some other causes of loss, like fire. *Hull* covers damage on the policyholder's own vehicle. The TPL insurance is mandatory—you are not allowed to drive without it—while the others are optional. The three types of cover are often sold in a package as a comprehensive insurance, but they are usually priced separately. In this chapter we will study partial casco only. Note that the concept *partial casco* is not used in all countries.

The Wasa tariff was based on three rating factors, as described in Table 1.1. For every vehicle, the data contains the exact moped age, but the choice was to have just two age classes. As for weight and gears the data does not contain any other information than the vehicle class, as defined in Table 1.1. The geographic areas, here called "zones", is a standard classification of Swedish parishes based on

Table 1.1 Rating factors in moped insurance

Rating factor	Class	Class description
Vehicle class	1	Weight over 60 kg and more than two gears
	2	Other
Vehicle age	1	At most 1 year
	2	2 years or more
Geographic zone	1	Central and semi-central parts of Sweden's three largest cities
	2	Suburbs and middle-sized towns
	3	Lesser towns, except those in 5 or 7
	4	Small towns and countryside, except 5–7
	5	Northern towns
	6	Northern countryside
	7	Gotland (Sweden's largest island)

different characteristics, as roughly described in the table. Obviously a large number of alternative groupings are possible.

The statistical study an actuary performs to obtain a tariff is called a *tariff analysis* here. It is based on the company's own data on insurance policies and claims for the portfolio, sometimes supplemented by external data, for instance from a national statistical agency. We next define some basic concepts used in connection with a tariff analysis.

The duration of a policy is the amount of time it is in force, usually measured in years, in which case one may also use the term *policy years*. The duration of a group of policies is obtained by adding the duration of the individual policies.

A claim is an event reported by the policy holder, for which she/he demands economic compensation.

The claim frequency is the number of claims divided by the duration, for some group of policies in force during a specific time period, i.e., the average number of claims per time period (usually one year). Claim frequencies are often given *per mille*, measuring the number of claims per one thousand policy years.

The claim severity is the total claim amount divided by the number of claims, i.e., the average cost per claim. Frequently, one omits *nil claims* (claims which turn out to yield no compensation from the insurer) from the claim frequency and severity—it is of course important to be consistent here and either omit or include nil claims in both cases.

The pure premium is the total claim amount divided by the duration, i.e., average cost per policy year (or other time period). Consequently, the pure premium is the product of the claim frequency and the claim severity.

Pure premium = Claim frequency × Claim severity

The earned premium is the amount of the premium income that covers risks during the period under study. Commonly, premiums are assumed to be earned *pro rata temporis*, i.e. linearly with time. This means that the earned premium is the duration, measured in years, times the annual premium.

The loss ratio is the claim amount divided by the earned premium. Related is the *combined ratio*, which is the ratio of the claim amount *plus* administrative expenses to the earned premium.

The claim frequency, claim severity, pure premium and loss ratio defined above are called *key ratios*. Let us have a look at some of these key ratios in the moped tariff of Example 1.1.

Table 1.2 Key ratios in moped insurance (claim frequency per mille)

Tariff cell			Duration	No.	Claim	Claim	Pure	Actual
Class	Age	Zone		claims	frequency	severity	premium	premium
1	1	1	62.9	17	270	18 256	4 936	2 049
1	1	2	112.9	7	62	13 632	845	1 230
1	1	3	133.1	9	68	20 877	1 411	762
1	1	4	376.6	7	19	13 045	242	396
1	1	5	9.4	0	0	.	0	990
1	1	6	70.8	1	14	15 000	212	594
1	1	7	4.4	1	228	8 018	1 829	396
1	2	1	352.1	52	148	8 232	1 216	1 229
1	2	2	840.1	69	82	7 418	609	738
1	2	3	1 378.3	75	54	7 318	398	457
1	2	4	5 505.3	136	25	6 922	171	238
1	2	5	114.1	2	18	11 131	195	594
1	2	6	810.9	14	17	5 970	103	356
1	2	7	62.3	1	16	6 500	104	238
2	1	1	191.6	43	224	7 754	1 740	1 024
2	1	2	237.3	34	143	6 933	993	615
2	1	3	162.4	11	68	4 402	298	381
2	1	4	446.5	8	18	8 214	147	198
2	1	5	13.2	0	0	.	0	495
2	1	6	82.8	3	36	5 830	211	297
2	1	7	14.5	0	0	.	0	198
2	2	1	844.8	94	111	4 728	526	614
2	2	2	1 296.0	99	76	4 252	325	369
2	2	3	1 214.9	37	30	4 212	128	229
2	2	4	3 740.7	56	15	3 846	58	119
2	2	5	109.4	4	37	3 925	144	297
2	2	6	404.7	5	12	5 280	65	178
2	2	7	66.3	1	15	7 795	118	119

Example 1.2 (Moped insurance, contd.) Table 1.2 is based on 38 508 policies and 860 associated claims during 1994–1999. Originally, the data contained some *nil claims*, which are excluded from the analysis here. The first three columns specify the tariff cell, characterized by the *vehicle class*, *vehicle age* and *zone* for the mopeds in that cell.

Let us compare the old tariff to the observed pure premium. "Actual premium" is the premium for one year according to the tariff in force 1999; it should be noted that this includes loadings for expenses, capital cost etc., and is thus not directly comparable to the column "pure premium", but the relative prices in two cells may

Table 1.3 Important key ratios

Exposure w	Response X	Key ratio $Y = X/w$
Duration	Number of claims	Claim frequency
Duration	Claim cost	Pure premium
Number of claims	Claim cost	(Average) Claim severity
Earned premium	Claim cost	Loss ratio
Number of claims	Number of large claims	Proportion of large claims

be compared. Note also that, despite its name, the pure premium is not a premium but only the average cost per annum over the years 1994–1999. We can see that cells with large annual premium tend to have large observed pure premium, too. At the same time, it appears that the general premium level is much too low in this case.

The key ratios are all of the same type—a ratio between the outcome of a random variable and a volume measure; the latter we call an *exposure*. The exposure w results in a response X, for instance the number of claims or the claim amount. The analysis is carried out on the key ratios $Y = X/w$, rather than on the responses themselves. The exposure plays a fundamental part in the analysis—the more exposure in the data we analyze, the less the variation of the key ratios. The models we shall study will have in common that they describe how a key ratio depends on the rating factors.

In the personal lines of business, with a large number of similar insured objects, such as cars and private houses, a tariff analysis is normally based on claim frequency and claim severity. In the commercial lines, the duration may not be available or useful; here one might analyze the loss ratio, with earned premium as exposure.

Table 1.3 summarizes some common key ratios. We have already discussed the first four of these. The *proportion of large claims* is the proportion of claims with cost above some threshold, which, unlike the other key ratios, is forced to lie between 0 and 1.

Note that the key ratios *claim severity* and *proportion of large claims* are considered conditional on the number of claims. While *number of claims* is a random variable when we consider *claim frequency*, it is thus a non-random exposure w in the two mentioned cases.

1.2 Basic Model Assumptions

We now present a set of basic assumptions that will give a foundation for our statistical model building.

Assumption 1.1 (Policy independence) Consider n different policies. For any response type in Table 1.3, let X_i denote the response for policy i. Then $X_1, \ldots, X_n$ are independent.

This is a fundamental assumption, but it is not hard to find examples where it is not fulfilled: in motor insurance, the possibility of a collision between two cars that are insured by the same company violates the independence assumption, at least in principle, but the effect of neglecting this should be small. A more important example of dependence is catastrophes, in which a large number of policies are affected by the same wind storm or flood, say. This situation calls for other types of models than those treated in this text and the setting is quite different: rating factors in our meaning are not used. Catastrophes are often singled out from the "normal" claims and special reinsurance cover is bought to reduce their impact. Relevant models can be found in any textbook in risk theory, e.g., Daykin et al. [DPP94] or Dickson [Di05]. An interesting mathematical model for wind storm losses is given by Rootzén and Tajvidi [RT97].

Assumption 1.2 (Time independence) Consider n disjoint time intervals. For any response type in Table 1.3, let X_i denote the response in time interval i. Then $X_1, \ldots, X_n$ are independent.

Here, both the number of claims and the claim amount are assumed independent from one time period to another. There are, of course, possible objections to this assumption as well: a car driver that has had an accident may drive more carefully in the future or a house owner that has experienced housebreaking might install a burglar alarm; in such cases, a claim implies a lower claim frequency in the future. In general, though, time independence seems a reasonable assumption that substantially simplifies the model building.

A consequence of the above two assumptions is that all individual claim costs are independent—they either concern different policies or occur in different time intervals.

The idea behind most existing tariffs is to use the rating factors to divide the portfolio of insurance policies into homogenous groups—the already mentioned tariff cells. It is then reasonable to charge the same premium for all policies within a tariff cell (for the same duration).

Assumption 1.3 (Homogeneity) Consider any two policies in the same tariff cell, having the same exposure. For any response type in Table 1.3, let X_i denote the response for policy i. Then X_1 and X_2 have the same probability distribution.

This homogeneity assumption is typically not perfectly fulfilled. In practice we look for a model that divides the policies into *fairly* homogeneous groups and charge the same premium within each tariff cell. To cope with residual non-homogeneity within cells, there are *bonus/malus systems* for private motor cars and *experience rating* of large businesses. This kind of individual rating can be obtained from the credibility models that we introduce in Chap. 4; until then, we work under Assumption 1.3.

Assumption 1.3 contains an implicit supposition of homogeneity over time: the two policies may be of the same length but be in force during different calendar periods; the second policy may even be a *renewal* of the first one, i.e., the policies are taken by the same customer for the same insured object. Assumption 1.3 implies that the only thing that matters is the duration of a policy, not when it starts or ends. This may appear unrealistic, since many claim types are subject to seasonal variation. Over larger periods of time, there may also be trends, in particular in claim costs (inflation). Notwithstanding this fact, homogeneity is essential in providing repeated observations for the statistical analysis, and the objections are overcome as follows.

In practice a tariff analysis usually targets the annual premium and data are, to a large extent, from one-year policies; hence, seasonal variation is not really a problem. As for trends, there are several alternatives to reduce their impact: recalculating claim amounts to current prices by some price index, using time as an explanatory variable (rating factor) in our models or use a short enough data capture period to make trend effects negligible. Furthermore, trends often mainly affects the overall premium level, while the relations between the tariff cells—which are the target of a tariff analysis—are more stable over time.

1.2.1 Means and Variances

A tariff analysis targets the *expected values* in the tariff cells, but to find the precision of our estimates we will also need some knowledge on the *variances*. So let us investigate the implications of Assumptions 1.1–1.3 on expectations and variances. Consider an arbitrary key ratio $Y = X/w$ for a group of policies in a tariff cell with total exposure w and total response X.

First consider the situation where w is the number of claims, so that we may write X as a sum of w individual responses $Z_1, \ldots, Z_w$. (If X is claim cost, then Z_k is the cost for the kth claim.) Assumptions 1.1–1.2 imply that the Z_k's are independent, since the claims come from different policies or different points in time; Assumption 1.3 implies identical distribution, so we may write $E(Z_k) = \mu$ and $\mathrm{Var}(Z_k) = \sigma^2$, for some μ and σ^2. Now elementary rules for expectation and variance directly yield

$$E(X) = w\mu, \qquad \mathrm{Var}(X) = w\sigma^2, \tag{1.1}$$

$$E(Y) = \mu, \qquad \mathrm{Var}(Y) = \sigma^2/w. \tag{1.2}$$

In the following lemma, we claim that theses results are valid also for key ratios where the exposure is *duration* or *earned premium*.

Lemma 1.1 *Under Assumptions* 1.1, 1.2 *and* 1.3, *if X is any response in Table* 1.3 *with $w > 0$ and $Y = X/w$, then the expectation and variance of X and Y are given by* (1.1) *and* (1.2), *respectively. Here μ and σ^2 is the expectation and the variance, respectively, for a response with exposure $w = 1$.*

Proof We have already seen that the result is valid when w is the number of claims. Consider now a w that is a rational number $w = m/n$. Then the exposure can be subdivided into m parts of $1/n$ each. In case w is *duration* this is done by taking m time intervals of equal length; when w is *earned premium* we make the time intervals long enough to get the premium $1/n$ in each, assuming that the premium is charged continuously in time.

The responses in these intervals are denoted by $Z_1, \ldots, Z_m$, which is a collection of independent and identically distributed random variables. If we add n such responses Z_k we get a variable Z with exposure $w = 1$; by Assumption 1.3 all such Z have the same expectation and variance and we may write $E(Z) = \mu$ and $\text{Var}(Z) = \sigma^2$. Now $E(Z) = nE(Z_1)$ and $\text{Var}(Z) = n\,\text{Var}(Z_1)$, yielding $E(Z_k) = E(Z_1) = \mu/n$ and $\text{Var}(Z_k) = \text{Var}(Z_1) = \sigma^2/n$. Since $X = \sum_{k=1}^{m} Z_k$ we find $E(X) = \mu m/n$ and $\text{Var}(X) = \sigma^2 m/n$, so that (1.1) is again valid, and (1.2) follows immediately.

This proves the lemma when the exposure is a rational number, which is of course always the case in practice. The transition to all real w requires taking the limits for an approximating sequence of rational numbers. We refrain form a strict proof here, since this would hardly contribute to the understanding of the result. □

A consequence of this lemma is that we consistently should use *weighted* variances in all models for key ratios.

1.3 Multiplicative Models

A tariff analysis is usually based on the insurer's own data. If we had enough claims data in each tariff cell, we could determine a premium for the cell by simply estimating the expected cost by the observed pure premium (corrected for inflation and loaded for expenses, etc.). In practice this is hardly ever the case—see for example Table 1.2 where some cells do not even have a single claim for the period 1994–1999. Hence, there is a need for methods giving an expected pure premium that varies more smoothly over the cells, with good precision of the cell estimates. It is also important that premiums are relatively stable over time and not subject to large random fluctuations.

This is achieved by using a model for how the expected pure premium depends on a number of rating factors. In this section we will describe the much used *multiplicative model*. Besides its use with the pure premium, a tariff can be produced by in turn applying it to claim frequency and severity.

Say that we have M rating factors, each one divided into classes, where m_i denotes the number of classes for rating factor i. For simplicity, we set out by considering the case with just two rating factors, i.e., $M = 2$. Then we can denote a tariff cell (i, j), where i and j denote the class of the first and second rating factor, respectively. In cell (i, j) we have the exposure w_{ij} and the response X_{ij}, leading to the key ratio $Y_{ij} = X_{ij}/w_{ij}$. According to Lemma 1.1 we have $E(Y_{ij}) = \mu_{ij}$, where

μ_{ij} is the expectation under unit exposure $w_{ij} = 1$. The multiplicative model is

$$\mu_{ij} = \gamma_0 \gamma_{1i} \gamma_{2j}. \tag{1.3}$$

Here $\{\gamma_{1i}; i = 1, \ldots, m_1\}$ are parameters that correspond to the different classes for rating factor number 1 and $\{\gamma_{2j}; j = 1, \ldots, m_2\}$ are the ones for factor 2, while γ_0 is a base value as will now be explained.

The model, as it stands, is over-parameterized: if we multiply all γ_{1i} by 10 and divide all γ_{2j} by 10, we get the same μ_{ij}'s as before. To make the parameters unique we specify a reference cell, called the base cell, preferably one with large exposure. Say for simplicity that the base cell is $(1, 1)$; then we put $\gamma_{11} = \gamma_{21} = 1$. Now γ_0 can be interpreted as a base value—the key ratio for the policies in the base cell—and the other parameters measure the relative difference in relation to the base cell and are called (price) *relativities*. For instance, if $\gamma_{12} = 1.25$ then the mean in cell $(2, 1)$ is 25% higher than in cell $(1, 1)$; the same is true for cell $(2, 2)$.

The multiplicativity assumption means that there is no interaction between the two rating factors. If rating factor number 1 is age class, number 2 is geographic region and the key ratio is pure premium, then an interpretation is that the relation of the pure premium for two age classes is the same within any region: if, say, the pure premium is 20% higher for ages 21–25 than for ages 26–30 in one region, then the same relation holds in any region. This is usually much more fair than to use an additive model. Say that the premium for ages 26–30 in region 1 is EUR 100, while in region 2 it is EUR 500. Then a multiplicative increase by 20% means adding EUR 20 and EUR 100, respectively, which might be fair if younger policyholders represent a larger risk to the company. If we should use an additive model, $\mu_{ij} = \gamma_0 + \gamma_{1i} + \gamma_{2j}$, the premium would increase by the same amount, e.g. EUR 20, in all regions, which makes a difference in region 1, but is almost negligible in relation to the total premium in region 2.

See Murphy, Brockman and Lee [MBL00] for further argumentation for the multiplicative model, which we will use almost exclusively from now on.

The simple multiplicative model in the case $M = 2$ can easily be extended to the general case, at the cost of somewhat cumbersome notation,

$$\mu_{i_1, i_2, \ldots, i_M} = \gamma_0 \gamma_{1i_1} \gamma_{2i_2} \cdots \gamma_{Mi_M}. \tag{1.4}$$

In Sect. 2.1, a more convenient notational system will be introduced.

As a final note, recall the distinction between determining the overall premium and the question of how to divide the required premium among the policies in the portfolio. The multiplicative model is ideally suited for handling this: the overall level is controlled by adjusting the base value γ_0, while the other parameters control how much to charge for a policy, given this base value. In practice the relativities γ_{ki_k} are determined first, and then the base value is set to give the required overall premium.

1.3.1 The Method of Marginal Totals

In the next chapter, we will show how the parameters in (1.3) and (1.4) can be estimated by generalized linear models (GLMs). To give some historical background, we shall present one of several earlier methods, *the method of marginal totals* (MMT), introduced by Bailey [Ba63]. The MMT is based on the following idea: even if we do not have enough data on the claim amount in each cell, we might have so on the margins, i.e. when we sum over all rating factors except one. Set this more stable marginal sum equal to its expectation and solve for the parameters, cf. the method of moments. In the simple case with only two rating factors this gives the following equations, using lower-case x for the outcome of the random variable X,

$$\sum_j E(X_{ij}) = \sum_j x_{ij}; \quad i = 1, \ldots, m_1;$$

$$\sum_i E(X_{ij}) = \sum_i x_{ij}; \quad j = 1, \ldots, m_2;$$

which under the multiplicative model in (1.3) yield

$$\begin{aligned} \sum_j w_{ij}\gamma_0\gamma_{1i}\gamma_{2j} &= \sum_j w_{ij}y_{ij}; \quad i = 1, \ldots, m_1; \\ \sum_i w_{ij}\gamma_0\gamma_{1i}\gamma_{2j} &= \sum_i w_{ij}y_{ij}; \quad j = 1, \ldots, m_2; \end{aligned} \tag{1.5}$$

where, nota bene, $\gamma_{11} = \gamma_{21} = 1$.

This system of equations has no solution in closed form—it has to be solved numerically. After some rearrangement and omittance of a redundant equation we can write the system of equations as follows.

$$\gamma_0 = \frac{\sum_i \sum_j w_{ij}y_{ij}}{\sum_i \sum_j w_{ij}\gamma_{1i}\gamma_{2j}}; \tag{1.6}$$

$$\gamma_{1i} = \frac{\sum_j w_{ij}y_{ij}}{\gamma_0 \sum_j w_{ij}\gamma_{2j}}; \quad i = 2, \ldots, m_1; \tag{1.7}$$

$$\gamma_{2j} = \frac{\sum_i w_{ij}y_{ij}}{\gamma_0 \sum_i w_{ij}\gamma_{1i}}; \quad j = 2, \ldots, m_2. \tag{1.8}$$

Starting with any positive numbers, the system can be solved by iterating over the above three equations.

It is straight-forward to extend these equations to cases with three or more variables.

Remark 1.1 For claim frequencies, Jung [Ju68] derived the MMT by maximum likelihood under a Poisson distribution.

Table 1.4 Tariff analysis for moped insurance: relativities according to the tariff and MMT (the method of marginal totals)

Rating factor	Class	Duration	Relativities in the tariff	Relativities of MMT
Vehicle class	1	9 833	1.00	1.00
	2	8 824	0.50	0.43
Vehicle age	1	1 918	1.66	2.73
	2	16 740	1.00	1.00
Zone	1	1 451	5.16	8.97
	2	2 486	3.10	4.19
	3	2 889	1.92	2.52
	4	10 069	1.00	1.00
	5	246	2.50	1.24
	6	1 369	1.50	0.74
	7	147	1.00	1.23

Example 1.3 (Moped insurance contd.) We apply the method of marginal totals (MMT) to the moped example, with the result shown in Table 1.4.

In the above example one may wonder whether the observed deviation of the estimated relativities from the current tariff are significant or just random? Some of the zones have rather few observations, and this could result in unstable estimates; but how many observations is enough to get reliable results? An advantage of the GLM approach over MMT and other early methods is that standard statistical theory provides tools for answering questions like this, in the form of confidence intervals and statistical tests.

1.3.2 One Factor at a Time?

It is important to note that, as in any regression method, the relativities in the multiplicative model should be interpreted *ceteris paribus* ("other things being equal"). This is sometimes stated as "holding everything else constant" or "controlling for the other variables".

Ceteris paribus:
- "All other things being equal";
- Regression coefficients measure the effect *ceteris paribus*, i.e. when all other variables are held constant.

Table 1.5 Number of policy years, lorry data

Annual	Vehicle age	
mileage	New	Old
Low	47 039	190 513
High	56 455	28 612

Table 1.6 Claim frequencies, lorry data

Annual	Vehicle age		
mileage	New	Old	Total
Low	0.033	0.025	0.026
High	0.067	0.049	0.061
Total	0.051	0.028	

The opposite would be the naïve analysis of examining one factor at a time: here the relativities are computed simply from the marginal sums for that variable. This method can lead to quite inappropriate estimates, as shown in the next example.

Example 1.4 (One-factor-at-a-time) As an illustration we consider some simple data from *Länsförsäkringar Alliance* concerning motor TPL claim frequencies for lorries. Table 1.5 shows the number of policy years split up by *annual mileage* (*low/high*) and *vehicle age* (*new/old*). In Table 1.6 the corresponding empirical claim frequencies are given.

Suppose we are considering a multiplicative model for this set of data, where *annual mileage* is the first factor. The tariff cell $(1, 1)$, i.e. *low/new*, is chosen as the reference cell. The *one-factor-at-a-time* estimator of the age-factor γ_{22} is then the ratio between the marginal claim frequency for vehicle age *old* and vehicle age *new*, i.e. $0.028/0.051 = 0.54$. Using the GLM estimator that will be derived in Sect. 2.3.1, using a multiplicative model so that the effect of *annual mileage* is taken into consideration when estimating the age-factor γ_{22}, we get 0.74; this is considerably larger than the naïve estimate. The reason that the latter underestimates the factor is that older lorries to a much larger extent have a lower annual mileage; hence their lower claim frequency is partly due to that they are less exposed to risk. But this effect is already captured by the mileage-factor: using the naïve estimate will thus severely underestimate the pure premium of the old lorries.

Exercises

1.1 (Section 1.3) The column *Actual premium* in Table 1.2 gives the current tariff. Is this tariff multiplicative?

1.2 (Section 1.3) Show how the system of equations (1.6)–(1.8) can be derived from (1.5).

1.3 (Section 1.3) The MMT estimates of the three parameters in Example 1.4 turn out to be $\gamma_0 = 0.03305$, $\gamma_{12} = 2.01231$ and $\gamma_{22} = 0.74288$.

Verify numerically that these estimates fulfill (1.6)–(1.8), by using a spreadsheet or a calculator.

Chapter 2
The Basics of Pricing with GLMs

As described in the previous section, the goal of a tariff analysis is to determine how one or more key ratios Y vary with a number of rating factors. This is reminiscent of analyzing how the dependent variable Y varies with the covariates (explanatory variables) x in a multiple linear regression. Linear regression, or the slightly larger general linear model, is not fully suitable for non-life insurance pricing, though, since: (i) it assumes normally distributed random errors, while the number of insurance claims follows a discrete probability distribution on the non-negative integers, and claim costs are non-negative and often skewed to the right; (ii) in linear models, the mean is a linear function of the covariates, while multiplicative models are usually more reasonable for pricing, cf. Sect. 1.3.

Generalized linear models (GLMs) is a rich class of statistical methods, which generalizes the ordinary linear models in two directions, each of which takes care of one of the above mentioned problems:

- *Probability distribution.* Instead of assuming the normal distribution, GLMs work with a general class of distributions, which contains a number of discrete and continuous distributions as special cases, in particular the normal, Poisson and gamma distributions.
- *Model for the mean.* In linear models the mean is a linear function of the covariates x. In GLMs some monotone transformation of the mean is a linear function of the x's, with the linear and multiplicative models as special cases.

These two generalization steps are discussed in Sects. 2.1 and 2.2, respectively.

GLM theory is quite recent—the basic ideas were introduced by Nelder and Wedderburn [NW72]. Already in the first 1983 edition of the standard reference by McCullagh and Nelder there is an example using motor insurance data; in the second edition [MN89] this example can be found in Sects. 8.4.1 and 12.8.3. But it was not until the second half of the 90's that the use of GLMs really started spreading, partly in response to the extended needs for tariff analysis due to the deregulation of the insurance markets in many countries. This process was facilitated by the publication of some influential papers by British actuaries, such as [BW92, Re94, HR96]; see also [MBL00], written for the US Casualty Actuarial Society a few years later.

E. Ohlsson, B. Johansson, *Non-Life Insurance Pricing with Generalized Linear Models*, EAA Lecture Notes,
DOI 10.1007/978-3-642-10791-7_2, © Springer-Verlag Berlin Heidelberg 2010

Some advantages of using GLMs over earlier methods for rate making are:

- GLMs constitute a general statistical theory, which has well established techniques for estimating standard errors, constructing confidence intervals, testing, model selection and other statistical features.
- GLMs are used in many areas of statistics, so that we can draw on developments both within and without of actuarial science.
- There is standard software for fitting GLMs that can easily be used for a tariff analysis, such as the SAS, GLIM, R or GenStat software packages.

In spite of the possibility to use standard software, many insurance companies use specialized commercial software for rate making that is provided by major consulting firms.

2.1 Exponential Dispersion Models

Here we describe the *exponential dispersion models* (EDMs) of GLMs, which generalize the normal distribution used in the linear models.

In our discussion of the multiplicative model in (1.3) and (1.4) we used a way of organizing the data with the observations, $y_{i_1,i_2,\ldots,i_K}$, having one index per rating factor. This is suitable for displaying the data in a table, especially in the two-dimensional case y_{ij}, and will therefore be called *tabular form*.

In our general presentation of GLMs, we rather assume that the data are organized on *list form*, with the n observations organized as a column vector $\mathbf{y}' = (y_1, y_2, \ldots, y_n)$. Besides the key ratio y_i, each row i of the list contains the exposure weight w_i of the tariff cell, as well as the values of the rating factors. The transition from tabular form to list form amounts to deciding on an order to display the tariff cells; a simple example is given in Table 2.1. As further illustration, consider again the moped insurance example, for which Table 1.2 contains an implicit list form; in Table 2.2 we repeat the part of that table that gives the list form for analysis of claim frequency.

List form corresponds to the way we organize the data in a data base, such as a SAS table. Tabular form, on the other hand, is useful for demonstrative purposes; hence, the reader should try to get used to both forms.

By the assumptions in Sect. 1.2, the variables $Y_1, \ldots, Y_n$ are independent, as required in general GLM theory. The probability distribution of an EDM is given

Table 2.1 Transition from tabular form to list form in a 2 by 2 case

Married	Male	Female
Yes	y_{11}	y_{12}
No	y_{21}	y_{22}

$\Longrightarrow$

i	Married	Gender	Observation
1	Yes	M	y_1
2	Yes	F	y_2
3	No	M	y_3
4	No	F	y_4

Table 2.2 Moped tariff on list form (claim frequency per mille)

Tariff cell	Covariates			Duration (exposure)	Claim frequency
	Class	Age	Zone		
i	x_{i1}	x_{i2}	x_{i3}	w_i	y_i
1	1	1	1	62.9	270
2	1	1	2	112.9	62
3	1	1	3	133.1	68
4	1	1	4	376.6	19
5	1	1	5	9.4	0
6	1	1	6	70.8	14
7	1	1	7	4.4	228
8	1	2	1	352.1	148
9	1	2	2	840.1	82
⋮	⋮	⋮	⋮	⋮	⋮
21	2	1	7	14.5	0
22	2	2	1	844.8	111
23	2	2	2	1 296.0	76
24	2	2	3	1 214.9	30
25	2	2	4	3 740.7	15
26	2	2	5	109.4	37
27	2	2	6	404.7	12
28	2	2	7	66.3	15

by the following frequency function, specializing to a probability density function in the continuous case and a probability mass function in the discrete case,

$$f_{Y_i}(y_i;\theta_i,\phi) = \exp\left\{\frac{y_i\theta_i - b(\theta_i)}{\phi/w_i} + c(y_i,\phi,w_i)\right\}. \tag{2.1}$$

Here θ_i is a parameter that is allowed to depend on i, while the *dispersion parameter* $\phi > 0$ is the same for all i. The so called *cumulant function* $b(\theta_i)$ is assumed twice continuously differentiable, with invertible first derivative. For every choice of such a function, we get a family of probability distributions, e.g. the normal, Poisson and gamma distributions, see Example 2.1 and (2.3) and (2.6) below. Given the choice of $b(\cdot)$, the distribution is completely specified by the parameters θ_i and ϕ. The function $c(\cdot,\cdot,\cdot)$, which does not depend on θ_i, is of little interest in GLM theory.

Of course, the expression above is only valid for the y_i that are possible outcomes of Y_i—the *support*; for other values of y_i we tacitly assume $f_{Y_i}(y_i) = 0$. Examples of support we will encounter is $(0,\infty)$, $(-\infty,\infty)$ and the non-negative integers.

Further technical restrictions are that $\phi > 0$, $w_i \geq 0$ and that the parameter space must be open, i.e., θ_i takes values in an open set, such as $0 < \theta_i < 1$ (while a closed set such as $0 \leq \theta_i \leq 1$ is not allowed).

An overview of the theory of the so defined exponential dispersion models can be found in Jörgensen [Jö97]. We will presuppose no knowledge of the theory for EDMs, and our interest in them is restricted only to the role they play in GLMs.

Remark 2.1 If ϕ was regarded as fixed, (2.1) would define a so called one-parameter exponential family, see, e.g., Lindgren [Li93, p. 188]. If, on the other hand, ϕ is unknown then we usually do *not* have a two-parameter exponential family, but we do have an EDM.

Example 2.1 (The normal distribution) Here we show that the normal distribution used in (weighted) linear models is a member of the EDM class; note, though, that the normal distribution is seldom used in the applications we have in mind. Nevertheless, assume for the moment that we have a normally distributed key ratio Y_i. The expectation of observation i is denoted μ_i, i.e., $\mu_i = E(Y_i)$. Lemma 1.1 shows that the variance must be w_i-weighted; the σ^2 of that lemma is assumed to be the same for all i in linear models. We conclude that $Y_i \sim N(\mu_i, \sigma^2/w_i)$, where w_i is the exposure. Then the frequency function is

$$f_{Y_i}(y_i) = \exp\left\{\frac{y_i\mu_i - \mu_i^2/2}{\sigma^2/w_i} + c(y_i, \sigma^2, w_i)\right\}, \tag{2.2}$$

were we have separated out the part of the density not depending on μ_i,

$$c(y_i, \sigma^2, w_i) = -\frac{1}{2}\left(\frac{w_i y_i^2}{\sigma^2} + \log(2\pi\sigma^2/w_i)\right).$$

This is an EDM with $\theta_i = \mu_i, \phi = \sigma^2$ and $b(\theta_i) = \theta_i^2/2$. Hence, the normal distribution used in (weighted) linear models is an EDM; the unweighted case is, of course, obtained by letting $w_i \equiv 1$.

2.1.1 Probability Distribution of the Claim Frequency

Let $N(t)$ be the number of claims for an individual policy during the time interval $[0, t]$, with $N(0) = 0$. The stochastic process $\{N(t); t \geq 0\}$ is called the *claims process*. Beard, Pentikäinen and Pesonen [BPP84, Appendix 4] show that under assumptions that are close to our Assumptions 1.2–1.3, plus an assumption that claims do not cluster, the claims process is a *Poisson process*. This motivates us to assume a Poisson distribution for the number of claims of an individual policy during any given period of time. By the independence of policies, Assumption 1.1, we get a Poisson distribution also at the aggregate level of all policies in a tariff cell.

So let X_i be the number of claims in a tariff cell with duration w_i and let μ_i denote the expectation when $w_i = 1$. Then by Lemma 1.1 we have $E(X_i) = w_i\mu_i$, and so X_i follows a Poisson distribution with frequency function

$$f_{X_i}(x_i; \mu_i) = e^{-w_i\mu_i}\frac{(w_i\mu_i)^{x_i}}{x_i!}, \quad x_i = 0, 1, 2, \ldots.$$

We are more interested in the distribution of the claim frequency $Y_i = X_i/w_i$; in the literature, this case is often (vaguely) referred to as Poisson, too, but since it is rather a transformation of that distribution we give it a special name, the *relative Poisson distribution*. The frequency function is, for y_i such that $w_i y_i$ is a non-negative integer,

$$f_{Y_i}(y_i;\mu_i) = P(Y_i = y_i) = P(X_i = w_i y_i) = \mathrm{e}^{-w_i\mu_i}\frac{(w_i\mu_i)^{w_i y_i}}{(w_i y_i)!}$$

$$= \exp\{w_i[y_i \log(\mu_i) - \mu_i] + c(y_i, w_i)\}, \tag{2.3}$$

where $c(y_i, w_i) = w_i y_i \log(w_i) - \log(w_i y_i!)$. This is an EDM, as can be seen by reparameterizing it through $\theta_i = \log(\mu_i)$,

$$f_{Y_i}(y_i;\theta_i) = \exp\{w_i(y_i\theta_i - \mathrm{e}^{\theta_i}) + c(y_i, w_i)\}.$$

This is of the form given in (2.1), with $\phi = 1$ and the cumulant function $b(\theta_i) = \mathrm{e}^{\theta_i}$. The parameter space is $\mu_i > 0$, i.e., the open set $-\infty < \theta_i < \infty$.

Remark 2.2 Is the Poisson distribution realistic? In practice, the homogeneity within cells is hard to achieve. The expected claim frequency μ_i of the Poisson process may vary with time, but this is not necessarily a problem since the number of claims during a year will still be Poisson distributed. A more serious problem is that there is often considerable variation left *between* policies within cells. This can be modeled by letting the risk parameter μ_i itself be the realization of a random variable. This leads to a so called *mixed* Poisson distribution, with larger variance than standard Poisson, see, e.g., [KPW04, Sect. 4.6.3]; such models often fit insurance data better than the standard Poisson. We will return to this problem in Sects. 3.4 and 3.5.

It is a reasonable requirement for a probabilistic model of a claim frequency to be *reproductive*, in the following sense. Suppose that we work under the assumption that the claim frequency in each cell has a relative Poisson distribution. If a tariff analysis shows that two cells have similar expectation, we might decide to merge them into just one cell. Then it would be very strange if we got another probability distribution in the new cell. Luckily, this problem will not arise—on the contrary, the relative Poisson distribution is reproduced on the aggregated level, as we shall now show.

Let Y_1 and Y_2 be the claim frequency in two cells with exposures w_1 and w_2, respectively, and let both follow a relative Poisson distribution with parameter μ. If we merge these two cells, the claim frequency in the new cell will be the weighted average

$$Y = \frac{w_1Y_1 + w_2Y_2}{w_1 + w_2}.$$

Since $w_1Y_1 + w_2Y_2$ is the sum of two independent Poisson distributed variables, it is itself Poisson distributed. Hence Y follows a relative Poisson distribution with exposure $w_1 + w_2$ and the parameter is, by elementary rules for the expectation of a linear expression, μ.

As indicated above, a parametric distribution that is closed under this type of averaging will be called *reproductive*, a concept coined by Jörgensen [Jö97]. In fact, this is a natural requirement for any key ratio; fortunately, we will see below in Theorem 2.2 that all EDMs are reproductive.

2.1.2 A Model for Claim Severity

We now turn to claim severity, and again we shall build a model for each tariff cell, but for the sake of simplicity, let us temporarily drop the index i. The exposure for claim severity, the number of claims, is then written w. Recall that in this analysis, we *condition* on the number of claims so that the exposure weight is non-random, as it should be. The idea is that we first analyze claim frequency with the number of claims as the outcome of a random variable; once this is done, we condition on the number of claims in analyzing claim severity. Here, the total claim cost in the cell is X and the claim severity $Y = X/w$.

In the previous section, we presented a plausible motivation for using the Poisson distribution, under the assumptions on independence and homogeneity. However, in the claim severity case it is not at all obvious which distribution we should assume for X. The distribution should be positive and skewed to the right, so the normal distribution is not suitable, but there are several other candidates that fulfill the requirements. However, the gamma distribution has become more or less a de facto standard in GLM analysis of claim severity, see, e.g., [MBL00, p. 10] or [BW92, Sect. 3]. As we will show in Sect. 2.3.3, the gamma assumption implies that the standard deviation is proportional to μ, i.e., we have a constant coefficient of variation; this seems quite plausible for claim severity. In Sect. 3.5 we will discuss the possibility of constructing estimators without assuming a specific distribution, starting from assumptions for the mean and variance structure only.

For the time being, we assume that the cost of an individual claim is gamma distributed; this is the case $w = 1$. One of several equivalent parameterizations is that with a so called *index parameter* $\alpha > 0$, a *scale parameter* $\beta > 0$, and the frequency function

$$f(x) = \frac{\beta^\alpha}{\Gamma(\alpha)} x^{\alpha-1} e^{-\beta x}; \quad x > 0. \tag{2.4}$$

We denote this distribution $G(\alpha, \beta)$ for short. It is well known that the expectation is α/β and the variance α/β^2, see Exercise 2.4. Furthermore, sums of independent gamma distributions with the same scale parameter β are gamma distributed with the same scale and an index parameter which is the sum of the individual α, see Exercise 2.11. So if X is the sum of w independent gamma distributed random variables, we conclude that $X \sim G(w\alpha, \beta)$. The frequency function for $Y = X/w$ is then

$$f_Y(y) = wf_X(wy) = \frac{(w\beta)^{w\alpha}}{\Gamma(w\alpha)} y^{w\alpha-1} e^{-w\beta y}; \quad y > 0,$$

and so $Y \sim G(w\alpha, w\beta)$ with expectation α/β. Before transforming this distribution to EDM form, it is instructive to re-parameterize it through $\mu = \alpha/\beta$ and $\phi = 1/\alpha$. In Exercise 2.5 we ask the reader to verify that the new parameter space is given by $\mu > 0$ and $\phi > 0$. The frequency function is

$$\begin{aligned} f_Y(y) = f_Y(y; \mu, \phi) &= \frac{1}{\Gamma(w/\phi)} \left(\frac{w}{\mu\phi} \right)^{w/\phi} y^{(w/\phi)-1} e^{-wy/(\mu\phi)} \\ &= \exp\left\{ \frac{-y/\mu - \log(\mu)}{\phi/w} + c(y, \phi, w) \right\}; \quad y > 0, \end{aligned} \tag{2.5}$$

where $c(y, \phi, w) = \log(wy/\phi)w/\phi - \log(y) - \log\Gamma(w/\phi)$. We have $E(Y) = w\alpha/(w\beta) = \mu$ and $\mathrm{Var}(Y) = w\alpha/(w\beta)^2 = \phi\mu^2/w$, which is consistent with Lemma 1.1.

To show that the gamma distribution is an EDM we finally change the first parameter in (2.5) to $\theta = -1/\mu$; the new parameter takes values in the open set $\theta < 0$. Returning to the notation with index i, the frequency function of the claim severity Y_i is

$$f_{Y_i}(y_i; \theta_i, \phi) = \exp\left\{ \frac{y_i\theta_i + \log(-\theta_i)}{\phi/w_i} + c(y_i, \phi, w_i) \right\}. \tag{2.6}$$

We conclude that the gamma distribution is an EDM with $b(\theta_i) = -\log(-\theta_i)$ and hence we can use it in a GLM.

Remark 2.3 By now, the reader might feel that in the right parameterization, any distribution is an EDM, but that is not the case; the log-normal distribution, e.g., can not be rearranged into an EDM.

2.1.3 Cumulant-Generating Function, Expectation and Variance

The *cumulant-generating function* is the logarithm of the moment-generating function; it is useful for computing the expectation and variance of Y_i, for finding the distribution of sums of independent random variables, and more. For simplicity, we once more refrain from writing out the subindex i for the time being. The *moment-generating function* of an EDM is defined as $M(t) = E(e^{tY})$, if this expectation is finite at least for real t in a neighborhood of zero. For continuous EDMs we find, by using (2.1),

$$\begin{aligned} E(e^{tY}) &= \int e^{ty} f_Y(y; \theta, \phi)\, dy \\ &= \int \exp\left\{ \frac{y(\theta + t\phi/w) - b(\theta)}{\phi/w} + c(y, \phi, w) \right\} dy \end{aligned}$$

$$= \exp\left\{\frac{b(\theta + t\phi/w) - b(\theta)}{\phi/w}\right\}$$
$$\times \int \exp\left\{\frac{y(\theta + t\phi/w) - b(\theta + t\phi/w)}{\phi/w} + c(y,\phi,w)\right\} dy. \quad (2.7)$$

Recall the assumption that the parameter space of an EDM must be open. It follows, at least for t in a neighborhood of 0, i.e., for $|t| < \delta$ for some $\delta > 0$, that $\theta + t\phi/w$ is in the parameter space. Thereby, the last integral equals one and the factor preceding it is the moment-generating function, which thus exists for $|t| < \delta$.

In the discrete case, we get the same expression by changing the integrals in (2.7) to sums. By taking logarithms, we conclude that the cumulant-generating function (CGF), denoted $\Psi(t)$, exists for any EDM and is given by

$$\Psi(t) = \frac{b(\theta + t\phi/w) - b(\theta)}{\phi/w}, \quad (2.8)$$

at least for t in some neighborhood of 0. This is the reason for the name *cumulant function* that we have already used for $b(\theta)$.

As the name suggests, the CGF can be used to derive the so called cumulants; this is done by differentiating and setting $t = 0$. The first cumulant is the expected value; the second cumulant is the variance; the reader who is not familiar with this result is invited to derive it from well-known results for moment-generating functions in Exercise 2.8.

We use the above property to derive the expected value of an EDM as follows, recalling that we have assumed that $b(\cdot)$ is twice differentiable,

$$\Psi'(t) = b'(\theta + t\phi/w); \qquad E(Y) = \Psi'(0) = b'(\theta).$$

The second cumulant, the variance, is given by

$$\Psi''(t) = b''(\theta + t\phi/w)\phi/w; \qquad \mathrm{Var}(Y) = \Psi''(0) = b''(\theta)\phi/w.$$

As a check, let us see what this yields in the case of a normal distribution: here $b(\theta) = \theta^2/2$, $b'(\theta) = \theta$ and $b''(\theta) = 1$, whence $E(Y) = \theta = \mu$ and $\mathrm{Var}(Y) = \phi/w = \sigma^2/w$ as it should be.

In general, it is more convenient to view the variance as a function of the mean μ. We have just seen that $\mu = E(Y) = b'(\theta)$, and since this is assumed to be an invertible function, we may insert the inverse relationship $\theta = b'^{-1}(\mu)$ into $b''(\theta)$ to get the so called *variance function* $v(\mu) \doteq b''(b'^{-1}(\mu))$. Now we can express $\mathrm{Var}(Y)$ as the product of the variance function $v(\mu)$ and a scaling and weighting factor ϕ/w. Here are some examples.

Example 2.2 (Variance functions) For the relative Poisson distribution in Sect. 2.1.1 we have $b(\theta_i) = \exp(\theta_i)$, by which $\mu_i = b'(\theta_i) = \exp(\theta_i)$. Furthermore $b''(\theta_i) = \exp(\theta_i) = \mu_i$, so that $v(\mu_i) = \mu_i$ and $\mathrm{Var}(Y) = \mu_i/w_i$, since $\phi = 1$. In the case $w_i = 1$ this is just the well-known result that the Poisson distribution has variance equal to the mean.

Table 2.3 Example of variance functions

Distribution	Normal	Poisson	Gamma	Binomial
$v(\mu)$	1	μ	μ^2	$\mu(1-\mu)$

The gamma distribution has $b(\theta_i) = -\log(-\theta_i)$ and $b'(\theta_i) = -1/\theta_i$ so that $\mu_i = -1/\theta_i$, as we already knew. Furthermore, $b''(\theta_i) = 1/\theta_i^2 = \mu_i^2$ so that $\mathrm{Var}(Y_i) = \phi\mu_i^2/w_i$.

The variance functions we have seen so far are collected in Table 2.3, together with the binomial distribution that is treated in Exercises 2.6 and 2.9.

We summarize the results in the following lemma, returning to our usual notation with index i for the observation number.

Lemma 2.1 *Suppose that Y_i follows an EDM, with frequency function given in (2.1). Then the cumulant generating function exists and is given by*

$$\Psi(t) = \frac{b(\theta_i + t\phi/w_i) - b(\theta_i)}{\phi/w_i},$$

and

$$\mu_i \doteq E(Y_i) = b'(\theta_i);$$
$$\mathrm{Var}(Y_i) = \phi v(\mu_i)/w_i,$$

where the variance function $v(\mu_i)$ is $b''(\cdot)$ expressed as a function of μ_i, i.e., $v(\mu_i) = b''(b'^{-1}(\mu_i))$.

Remark 2.4 Recall that in linear regression, we have a constant variance $\mathrm{Var}(Y_i) = \phi$, plus possibly a weight w_i which we disregard for a moment. The most general assumption would be to allow $\mathrm{Var}(Y_i) = \phi_i$, but this would make the model heavily over-parameterized. In GLMs we are somewhere in-between these extremes, since the variance function $v(\mu_i)$ allows the variance to vary over the cells i, but without introducing any new parameters.

The variance function is important in GLM model building, a fact that is emphasized by the following theorem.

Theorem 2.1 *Within the EDM class, a family of probability distributions is uniquely characterized by its variance function.*

The practical implication of this theorem is that if you have decided to use a GLM, and hence an EDM, you only have to determine the variance function; then you know the precise probability distribution within the EDM class. This is an interesting result: only having to model the mean and variance is much simpler than having to specify an entire distribution.

The core in the proof of this theorem is to notice that since $v(\cdot)$ is a function of the derivatives of $b(\cdot)$, the latter can be determined from $v(\cdot)$ by solving a pair of differential equations—but $b(\cdot)$ is all we need to specify the EDM distribution in (2.1). The proof can be found in Jörgensen [Jö87, Theorem 1].

We saw in Sect. 2.1.1 that the relative Poisson distribution had the appealing property of being reproductive. The next theorem shows that this holds for all distributions within the EDM class.

Theorem 2.2 (EDMs are reproductive) *Suppose we have two independent random variables Y_1 and Y_2 from the same EDM family, i.e., with the same $b(\cdot)$, that have the same mean μ and dispersion parameter ϕ, but possibly different weights w_1 and w_2. Then their w-weighted average $Y = (w_1Y_1 + w_2Y_2)/(w_1 + w_2)$ belongs to the same EDM distribution, but with weight $w. = w_1 + w_2$.*

The proof is left as Exercise 2.12, using some basic properties of CGFs derived in Exercise 2.10. From an applied point of view the importance of this theorem is that if we merge two tariff cells, with good reason to assume they have the same mean, we will stay within the same family of distributions. From this point of view, EDMs behave the way we want a probability distribution in pricing to behave. One may show that the log-normal and the (generalized) Pareto distribution do not have this property; our conclusion is that, useful as they may be in other actuarial applications, they are not well suited for use in a tariff analysis.

2.1.4 Tweedie Models

In non-life actuarial applications, it is often desirable to work with probability distributions that are closed with respect to scale transformations, or *scale invariant*. Let c be a positive constant, $c > 0$, and Y a random variable from a certain family of distributions; we say that this family is scale invariant if cY follows a distribution in the same family. This property is desirable if Y is measured in a monetary unit: if we convert the data from one currency to another, we want to stay within the same family of distributions—the result of a tariff analysis should not depend on the currency used. Similarly, the inference should not depend on whether we measure the claim frequency in per cent or per mille. We conclude that scale invariance is desirable for all the key ratios in Table 1.3, except for the proportion of large claims, for which scale is not relevant.

It can be shown that the only EDMs that are scale invariant are the so called Tweedie models, which are defined as having variance function

$$v(\mu) = \mu^p \tag{2.9}$$

for some p. The proof can be found in Jörgensen [Jö97, Chap. 4], upon which much of the present section is based, without further reference.

Table 2.4 Overview of Tweedie models

	Type	Name	Key ratio
$p<0$	Continuous	–	–
$p=0$	Continuous	Normal	–
$0<p<1$	Non-existing	–	–
$p=1$	Discrete	Poisson	Claim frequency
$1<p<2$	Mixed, non-negative	Compound Poisson	Pure premium
$p=2$	Continuous, positive	Gamma	Claim severity
$2<p<3$	Continuous, positive	–	Claim severity
$p=3$	Continuous, positive	Inverse normal	Claim severity
$p>3$	Continuous, positive	–	Claim severity

With the notable exception of the relative binomial in Exercise 2.6, for which scale invariance is not required since y is a proportion, the EDMs in this book are all Tweedie models. In Table 2.4 we give a list of all Tweedie models and the key ratios for which they might be useful. The cases $p = 0, 1, 2$ have already been discussed. Tweedie models with $p \geq 2$ are often suggested as distributions for the claim severity, especially $p = 2$, but also the *inverse normal distribution* ($p = 3$) is sometimes mentioned in the literature.

The class of models with $1 < p < 2$ is interesting: these so called *compound Poisson distributions* arise as the distribution of a sum of a Poisson distributed number of claims which follow a gamma distribution; hence, they are proper for modeling the pure premium, without doing a separate analysis of claim frequency and severity. Note that the compound Poisson is mixed (neither purely discrete nor purely continuous), having positive probability at zero plus a continuous distribution on the positive real numbers.

Jörgensen and Souza [JS94] analyze the pure premium for a private motor insurance portfolio in Brazil, and get the value $p = 1.37$ from an algorithm they designed for maximum likelihood estimation of p.

A bit surprising is that for $0 < p < 1$ no EDM exists. Negative values of p, finally, are allowed and give continuous distributions on the whole real axis, but to the best of our knowledge no application in insurance has been proposed.

From now on, we only discuss the case $p \geq 1$, which covers our applications, and we start by presenting the corresponding cumulant function $b(\theta)$.

$$b(\theta) = \begin{cases} e^{\theta}, & \text{for } p = 1; \\ -\log(-\theta), & \text{for } p = 2; \\ -\frac{1}{p-2}[-(p-1)\theta]^{(p-2)/(p-1)}, & \text{for } 1 < p < 2 \text{ and } p > 2. \end{cases} \tag{2.10}$$

The parameter space M_θ is

$$M_\theta = \begin{cases} -\infty < \theta < \infty, & \text{for } p = 1; \\ -\infty < \theta < 0, & \text{for } p > 1. \end{cases} \tag{2.11}$$

The derivative $b'(\theta)$ is given by

$$b'(\theta) = \begin{cases} e^{\theta}, & \text{for } p = 1; \\ [-(p-1)\theta]^{-1/(p-1)}, & \text{for } p > 1, \end{cases} \tag{2.12}$$

with inverse

$$h(\mu) = \begin{cases} \log(\mu), & p = 1; \\ -\frac{1}{p-1}\mu^{-(p-1)}, & p > 1. \end{cases} \tag{2.13}$$

These results are taken from Jörgensen [Jö97]; some are also verified in Exercise 2.13.

2.2 The Link Function

We have seen how to generalize the normal error distribution to the EDM class, and now turn to the other generalization of ordinary linear models, concerning the linear structure of the mean.

We start by discussing a simple example, in which we only have two rating factors, one with two classes and one with three classes. On tabular form, we let μ_{ij} denote the expectation of the key ratio in cell (i, j), where the first factor is in class i and the second is in class j. Linear models assume an *additive model* structure for the mean:

$$\mu_{ij} = \gamma_0 + \gamma_{1i} + \gamma_{2j}. \tag{2.14}$$

We recognize this model from the analysis of variance (ANOVA), and recall that it is over-parameterized, unless we add some restrictions. In an ANOVA the usual restriction is that marginal sums should be zero, but here we chose the restriction to force the parameters of some *base cell* to be zero. Say that $(1, 1)$ is the base cell; then we let $\gamma_{11} = \gamma_{21} = 0$, so that $\mu_{11} = \gamma_0$, and the other parameters measure the mean departure from this cell. Next we rewrite the model on list form by sorting the cells in the order $(1, 1)$; $(1, 2)$; $(1, 3)$; $(2, 1)$; $(2, 2)$; $(2, 3)$ and renaming the parameters,

$$\beta_1 \doteq \gamma_0,$$
$$\beta_2 \doteq \gamma_{12},$$
$$\beta_3 \doteq \gamma_{22},$$
$$\beta_4 \doteq \gamma_{23}.$$

With these parameters the expected values in the cells are as listed in Table 2.5.

Next, we introduce so called *dummy variables* through the relation

$$x_{ij} = \begin{cases} 1, & \text{if } \beta_j \text{ is included in } \mu_i, \\ 0, & \text{else.} \end{cases}$$

Table 2.5 Parameterization of a two-way additive model on list form

i	Tariff cell		μ_i			
1	1	1	β_1			
2	1	2	β_1		$+\beta_3$	
3	1	3	β_1			$+\beta_4$
4	2	1	β_1	$+\beta_2$		
5	2	2	β_1	$+\beta_2$	$+\beta_3$	
6	2	3	β_1	$+\beta_2$		$+\beta_4$

Table 2.6 Dummy variables in a two-way additive model

i	Tariff cell		x_{i1}	x_{i2}	x_{i3}	x_{i4}
1	1	1	1	0	0	0
2	1	2	1	0	1	0
3	1	3	1	0	0	1
4	2	1	1	1	0	0
5	2	2	1	1	1	0
6	2	3	1	1	0	1

The values of the dummy variables in this example are given in Table 2.6. Note the similarity to Table 2.5.

With these variables, the linear model for the mean can be rewritten

$$\mu_i = \sum_{j=1}^{4} x_{ij}\beta_j \quad i = 1, 2, \ldots, 6, \tag{2.15}$$

or on matrix form $\boldsymbol{\mu} = \mathbf{X}\boldsymbol{\beta}$, where $\mathbf{X}$ is called the *design matrix*, or *model matrix*, and

$$\boldsymbol{\mu} = \begin{pmatrix} \mu_1 \\ \mu_2 \\ \mu_3 \\ \mu_4 \\ \mu_5 \\ \mu_6 \end{pmatrix}, \qquad \mathbf{X} = \begin{pmatrix} x_{11} & x_{12} & x_{13} & x_{14} \\ x_{21} & x_{22} & x_{23} & x_{24} \\ x_{31} & x_{32} & x_{33} & x_{34} \\ x_{41} & x_{42} & x_{43} & x_{44} \\ x_{51} & x_{52} & x_{53} & x_{54} \\ x_{61} & x_{62} & x_{63} & x_{64} \end{pmatrix}, \qquad \boldsymbol{\beta} = \begin{pmatrix} \beta_1 \\ \beta_2 \\ \beta_3 \\ \beta_4 \end{pmatrix}. \tag{2.16}$$

So far the additive model for the mean has been used; next we turn to the multiplicative model $\mu_{ij} = \gamma_0\gamma_{1i}\gamma_{2j}$ that was introduced in Sect. 1.3. By taking logarithms we get

$$\log(\mu_{ij}) = \log(\gamma_0) + \log(\gamma_{1i}) + \log(\gamma_{2j}), \tag{2.17}$$

and again we must select a base cell, say (1, 1), where $\gamma_{11} = \gamma_{21} = 1$. Let us now do a transition to list form similar to the one we just performed for the additive model, by first letting

$$\begin{aligned}\beta_1 &\doteq \log \gamma_0,\\ \beta_2 &\doteq \log \gamma_{12},\\ \beta_3 &\doteq \log \gamma_{22},\\ \beta_4 &\doteq \log \gamma_{23}.\end{aligned}$$

By the aid of the dummy variables in Table 2.6, we have

$$\log(\mu_i) = \sum_{j=1}^{4} x_{ij}\beta_j; \quad i = 1, 2, \ldots, 6. \tag{2.18}$$

This is the same linear structure as in (2.15), the only difference being that the left-hand side is $\log(\mu_i)$ instead of just μ_i. In general GLMs this is further generalized to allow the left-hand side to be any monotone function $g(\cdot)$ of μ_i.

Leaving the simple two-way example behind, the general tariff analysis problem is to investigate how the response Y_i is influenced by r covariates $x_1, x_2, \ldots, x_r$. Introduce

$$\eta_i = \sum_{j=1}^{r} x_{ij}\beta_j; \quad i = 1, 2, \ldots, n, \tag{2.19}$$

where x_{ij} as before is the value of the covariate x_j for observation i.

In the ordinary linear model, $\mu_i \equiv \eta_i$; in a GLM this is generalized to an arbitrary relation $g(\mu_i) = \eta_i$, with the restriction that $g(\cdot)$ must be a monotone, differentiable function. This fundamental object in a GLM is called the *link function*, since it links the mean to the linear structure through

$$g(\mu_i) = \eta_i = \sum_{j=1}^{r} x_{ij}\beta_j. \tag{2.20}$$

We have seen that multiplicative models correspond to a logarithmic link function, a *log link*,

$$g(\mu_i) = \log(\mu_i),$$

while the linear model uses the *identity link* $\mu_i = \eta_i$, i.e., $g(\mu_i) = \mu_i$. Note that the link function is not allowed to depend on i.

For the analysis of proportions, it is common to use a *logit link*,

$$\eta_i = g(\mu_i) = \log\left(\frac{\mu_i}{1-\mu_i}\right). \tag{2.21}$$

This link guarantees that the mean will stay between zero and one, as required in a model where μ_i is a proportion, such as the relative binomial model for the key ratio *proportion of large claims*. The corresponding GLM analysis goes under the name *logistic regression*.

In a GLM, the link function is part of the model specification. In non-life insurance pricing, the log link is by far the most common one, since a multiplicative model is often reasonable. For a discussion of the merits of the multiplicative model, see Sect. 1.3.

Interactions may cause departure from multiplicativity. An example is motor insurance, where young men are more accident-prone than young women, while in midlife there is little gender difference in driving behavior. In Sect. 3.6.2 we will show how this problem can be handled within the general multiplicative framework. For further discussion on multiplicative models, see [BW92, Sects. 2.1 and 3.1.1]. With the exception of the analysis of proportions, we will assume a multiplicative model throughout this text, with possible interactions handled by combining variables as in the age and gender example.

We end this section by summarizing the GLM generalization of the ordinary linear model and the particular case that is most used in our applications.

Weighted linear regression models:
- Y_i follows a normal distribution with $\mathrm{Var}(Y_i) = \sigma^2/w_i$;
- The mean follows the additive model $\mu_i = \sum_j x_{ij}\beta_j$.

Generalized Linear Models (GLMs):
- Y_i follows an EDM with $\mathrm{Var}(Y_i) = \phi v(\mu_i)/w_i$;
- The mean satisfies $g(\mu_i) = \sum_j x_{ij}\beta_j$, where g is a monotone function.

Multiplicative Tweedie model, subclass of GLMs:
- Y_i follows a Tweedie EDM with $\mathrm{Var}(Y_i) = \phi\mu_i^p/w_i$, $p \geq 1$;
- The mean follows the multiplicative model $\log(\mu_i) = \sum_j x_{ij}\beta_j$.

2.2.1 Canonical Link*

In the GLM literature one encounters the concept of a *canonical link*; this concept is not so important in our applications, but for the sake of completeness we shall give a brief orientation on this subject. First we note that we have been working with several different parameterizations of a GLM; these parameters are unique functions of each other as illustrated in the following figure:

$$\theta \xrightarrow{b'(\cdot)} \mu \xrightarrow{g(\cdot)} \eta. \tag{2.22}$$

Here $b(\cdot)$, and hence $b'(\cdot)$, are determined by the structure of the random component, uniquely determined by the choice of variance function. The link function $g(\cdot)$, on the other hand, is part of the modeling of the mean and in some applications there are several reasonable choices. The special choice $g(\cdot) = b'^{-1}(\cdot)$ is the *canonical link*. From (2.22) we find that the canonical link makes $\theta = \eta$. It turns out that using the canonical link simplifies some computations, but the name is somewhat misleading since it may not be the natural choice in a particular application. On the other hand, some of the most common models actually use canonical links.

Example 2.3 (Some canonical links) The normal distribution has $\mu = b'(\theta) = \theta$ and the identity link $g(\mu) = \mu$ that is used in the linear model is the canonical one.

In the Poisson case we have $\mu = b'(\theta) = e^{\theta}$, by which the log link $g(\mu) = \log \mu$ is canonical. So for this important EDM, the multiplicative model is canonical.

In the case of the gamma distribution $b'(\theta) = -1/\theta$ and so the canonical link is the inverse link $g(\mu) = -1/\mu$. A problem with this link is that, as opposed to the log link, it may cause the mean to take on negative values. McCullagh and Nelder [MN89, Sects. 8.4.1 and 12.8.3], suggests using this link for claim severity, but as far as we know it is not used in practice.

2.3 Parameter Estimation

So far, we have defined GLMs and studied some of their properties. It is now time for the most important step: the estimation of the regression parameters in (2.20), from which we will get the relativities—the basic building blocks of the tariff. Before deriving a general result, we will give an introduction to the subject by treating an important special case.

2.3.1 The Multiplicative Poisson Model

We return to the simple case in Sect. 1.3.1, with just two rating factors. For easy reference, we repeat the multiplicative model on tabular form from (1.3)

$$\mu_{ij} = \gamma_0 \gamma_{1i} \gamma_{2j}. \tag{2.23}$$

In GLM terms, we say that we use a log link. In addition to this, we assume that the claim frequency Y_{ij} has a relative Poisson distribution, with frequency function given on list form in (2.3). We rewrite this on tabular form and note that the cells are independent due to Assumption 1.1, implying that the log-likelihood of the whole sample is the sum of the log-likelihoods in the individual cells.

$$\ell = \sum_i \sum_j w_{ij}\{y_{ij}[\log(\gamma_0) + \log(\gamma_{1i}) + \log(\gamma_{2j})] - \gamma_0\gamma_{1i}\gamma_{2j}\} + c, \tag{2.24}$$

where c does not depend on the γ parameters. By in turn differentiating ℓ with respect to each γ, we get a system of equations for the maximum likelihood estimates (MLEs), the so called ML equations. The result is exactly the MMT equations in (1.6)–(1.8); this is no surprise if we recall that Jung [Ju68] derived these equations as the MLEs of a Poisson model, cf. Remark 1.1.

In general, under a GLM model for the claim frequency with a relative Poisson distribution and log link (multiplicative model), the ML equations are equal to the equations of the method of marginal totals (MMT), and hence the resulting estimates are the same, see Exercise 2.16.

2.3.2 General Result

We now turn to the general case of finding the MLEs of the $\boldsymbol{\beta}$-parameters in a GLM. The estimates are based on a sample of n observations. The individual observations follow the EDM distribution given on list form in (2.1) and by independence, the log-likelihood as a function of $\boldsymbol{\theta}$ is

$$\ell(\boldsymbol{\theta};\phi,\mathbf{y}) = \frac{1}{\phi}\sum_i w_i(y_i\theta_i - b(\theta_i)) + \sum_i c(y_i,\phi,w_i). \tag{2.25}$$

It is clear that the dispersion parameter ϕ does not affect the maximization of ℓ with respect to $\boldsymbol{\theta}$—a similar observation should be familiar from the linear regression model, where ϕ is denoted σ^2. We can therefore ignore ϕ here, but we will return to the question of how to estimate it later, in Sect. 3.1.1.

The likelihood as a function of $\boldsymbol{\beta}$, rather than $\boldsymbol{\theta}$, could be found by the inverse of the relation $\mu_i = b'(\theta_i)$, combined with the link $g(\mu_i) = \eta_i = \sum_j x_{ij}\beta_j$. The derivative of ℓ with respect to β_j is, by the chain rule,

$$\begin{aligned}\frac{\partial \ell}{\partial \beta_j} &= \sum_i \frac{\partial \ell}{\partial \theta_i}\frac{\partial \theta_i}{\partial \beta_j} = \frac{1}{\phi}\sum_i \left(w_i y_i - w_i b'(\theta_i)\right)\frac{\partial \theta_i}{\partial \beta_j} \\ &= \frac{1}{\phi}\sum_i \left(w_i y_i - w_i b'(\theta_i)\right)\frac{\partial \theta_i}{\partial \mu_i}\frac{\partial \mu_i}{\partial \eta_i}\frac{\partial \eta_i}{\partial \beta_j}.\end{aligned} \tag{2.26}$$

By the mentioned relation $\mu_i = b'(\theta_i)$ we have $\partial\mu_i/\partial\theta_i = b''(\theta_i)$. The derivative of the inverse relation is simply the inverse of the derivative, and so $\partial\theta_i/\partial\mu_i = 1/v(\mu_i)$, since by definition $v(\mu_i) = b''(\theta_i)$.

Furthermore, $\partial\mu_i/\partial\eta_i = [\partial\eta_i/\partial\mu_i]^{-1} = 1/g'(\mu_i)$. Finally, from $\eta_i = \sum_j x_{ij}\beta_j$ we get $\partial\eta_i/\partial\beta_j = x_{ij}$.

Putting all this together gives the result,

$$\frac{\partial \ell}{\partial \beta_j} = \frac{1}{\phi}\sum_i w_i \frac{y_i - \mu_i}{v(\mu_i)g'(\mu_i)} x_{ij}, \tag{2.27}$$

the so called *score function*. By setting all these r partial derivatives equal to zero and multiplying by ϕ, we get the ML equations

$$\sum_i w_i \frac{y_i - \mu_i}{v(\mu_i)g'(\mu_i)} x_{ij} = 0, \quad j = 1, 2, \ldots, r. \tag{2.28}$$

It might look as if the solution is simply $\mu_i = y_i$, but then we forget that $\mu_i = \mu_i(\boldsymbol{\beta})$ also has to satisfy the relation given by the regression on the x's, i.e.,

$$\mu_i = g^{-1}(\eta_i) = g^{-1}\Big(\sum_j x_{ij}\beta_j\Big). \tag{2.29}$$

It is only the so called *saturated model*, where there are as many parameters as there are observations, that allows the solution $\mu_i = y_i$.

It is interesting to note that the only property of the probability distribution that affects the ML equations (2.28) is the mean and the variance, through the link function g and the variance function v; for further discussion on this issue, see Sect. 3.5.

Example 2.4 (Tweedie models) In a tariff analysis, we typically use the Tweedie models of Sect. 2.1.4, where $v(\mu) = \mu^p$, in connection with a multiplicative model for the mean, i.e., $g(\mu_i) = \log(\mu_i)$, implying that $g'(\mu_i) = 1/\mu_i$. Then the general ML equations in (2.28) simplify to

$$\sum_i w_i \frac{y_i - \mu_i}{\mu_i^{p-1}} x_{ij} = 0 \quad \Longleftrightarrow \quad \sum_i \frac{w_i}{\mu_i^{p-1}} y_i x_{ij} = \sum_i \frac{w_i}{\mu_i^{p-1}} \mu_i x_{ij}, \tag{2.30}$$

where the μ's are connected to the β's through

$$\mu_i = \exp\Big\{\sum_j x_{ij}\beta_j\Big\}. \tag{2.31}$$

Compared to the Poisson case $p = 1$, models with $p > 1$ will downweigh both sides of the right-most equation in (2.30) by μ^{p-1}, giving less weight to cells with large expectation.

In the end, we are not interested in the β's, but rather the relativities γ. These are found by the relation $\gamma_j = \exp\{\beta_j\}$.

Introduce the diagonal matrix $\mathbf{W}$ with the parameter-dependent "weights"

$$\tilde{w}_i = \frac{w_i}{v(\mu_i)g'(\mu_i)},$$

on the diagonal and zeroes off the diagonal, $\mathbf{W} = \operatorname{diag}(\tilde{w}_i; i = 1, \ldots, n)$. Then (2.28) may be written on matrix form

$$\mathbf{X}'\mathbf{W}\mathbf{y} = \mathbf{X}'\mathbf{W}\boldsymbol{\mu}, \tag{2.32}$$

where $\mathbf{X}$ is the design matrix, cf. the example in (2.16). In the case of the normal distribution with identity link, it is readily seen that $\tilde{w}_i = w_i$, and if we furthermore let all weights $w_i = 1$, then (2.32) is reduced to the well-known "normal equations" of the linear model $\mathbf{X}'\mathbf{y} = \mathbf{X}'\mathbf{X}\boldsymbol{\beta}$, as we would expect since the linear model is a special case of the GLM. Here we have used the fact that $\boldsymbol{\mu} = \mathbf{X}\boldsymbol{\beta}$ when we use the identity link.

Except for a few special cases, the ML equations must be solved numerically. The interested reader is referred to Sect. 3.2.3 for an introduction to numerical methods for solving the ML equations and to Sect. 3.2.4 for the question whether ML equations really give a (unique) maximum of the likelihood.

2.3.3 Multiplicative Gamma Model for Claim Severity

In Sect. 2.3.1 we discussed estimation for claim frequency; our next important special case is claim severity. With the data on list form, in cell i we have w_i claims and the claim severity Y_i, which is assumed to be relative gamma distributed with density given by (2.5). Let us have a closer look at the relation between the mean μ_i and the variance in this case. By Lemma 2.1 and Table 2.3 we have $E(Y_i) = \mu_i$ and $\mathrm{Var}(Y_i) = \phi\mu_i^2/w_i$. Hence,

$$\frac{\mathrm{Var}(Y_i)}{E(Y_i)^2} = \frac{\phi}{w_i}, \tag{2.33}$$

which means that the coefficient of variation (*CV*) is constant over cells with the same exposure w_i. This is, of course, a consequence of having a constant CV in the underlying model for *individual* claims. Such a constant CV, i.e., a standard deviation proportional to the mean, is plausible and in any case much more realistic than having a constant standard deviation: say that we have a tariff cell with mean 20 and standard deviation 4, then in another cell with the same exposure but with mean 200 we would rather expect a standard deviation of 40 than of 4.

From (2.30) with $p = 2$ we get the ML equations, which in this case are

$$\sum_i w_i \frac{y_i}{\mu_i} x_{ij} = \sum_i w_i x_{ij}. \tag{2.34}$$

It is instructive to write these equations on *tabular form* in the simple case with just two rating factors, using the multiplicative model in (2.23):

$$\begin{aligned}
\sum_i \sum_j \frac{w_{ij} y_{ij}}{\gamma_0 \gamma_{1i} \gamma_{2j}} &= \sum_i \sum_j w_{ij}; \\
\sum_j \frac{w_{ij} y_{ij}}{\gamma_0 \gamma_{1i} \gamma_{2j}} &= \sum_j w_{ij} \quad i = 2, \ldots, k_1; \\
\sum_i \frac{w_{ij} y_{ij}}{\gamma_0 \gamma_{1i} \gamma_{2j}} &= \sum_i w_{ij} \quad j = 2, \ldots, k_2.
\end{aligned} \tag{2.35}$$

This system of equations provides a quite natural algorithm for estimating relativities, giving a marginal balance in the sum of the relative deviations of the observations from their means: the average relative deviation equals one for each rating factor level i or j. Mack [Ma91] refers to work by Eeghen, Nijssen and Ruygt from 1982 where these equations are used for the pure premium, under the name *the direct method*. They found that the estimates of this method were very close to those given by the MMT. This is consistent with our general experience that the choice of p in the Tweedie models is not that important for estimating relativities.

2.3.4 Modeling the Pure Premium

In the end, it is the model for the pure premium that gives the tariff. One might consider using a Tweedie model with $1 < p < 2$ to analyze the pure premium directly, as demonstrated by Jörgensen and Souza [JS94]. However, the standard GLM tariff analysis is to do separate analyses for claim frequency and claim severity, and then relativities for the pure premium are found by multiplying the results. The reason for this separation into two GLMs is:

(i) Claim frequency is usually much more stable than claim severity and often much of the power of rating factors is related to claim frequency: these factors can then be estimated with greater accuracy;
(ii) A separate analysis gives more insight into how a rating factor affects the pure premium.

See also [BW92, MBL00].

Example 2.5 (Moped insurance contd.) We return once more to Example 1.1, with the aggregated data given in Table 1.2. In Example 1.3 we applied the MMT directly to the pure premium. We now carry out a separate analysis for claim frequency and severity, obtaining the relativities for the pure premium by multiplying the factors from these two analyses. The results are listed in Table 2.7.

We observe that the two variables *vehicle class* and *vehicle age* affect claim frequency and severity in the same direction, which means that newer and stronger vehicles are not only more expensive to replace when stolen, but are also stolen more often. The geographic zone has a large impact on the claim frequency, but once a moped is stolen the cost of replacing it is not necessarily larger in one zone than another, with a possible exception for the largest cities in zone 1. Note that these interesting conclusions could not have been drawn, had we analyzed pure premium only. On the other hand, the resulting estimates for pure premium are quite close to those obtained by the MMT method, see Table 1.4.

For zone 5 and 7, it is rather obvious that no conclusion can be drawn due to the very small number of claims, and hence very uncertain estimates of claim severity. But how can we know if the 23 observations in zone 6 or the 132 in zone 3 are enough to draw significant conclusions? This is one of the themes in the next chapter, where we go further into GLM theory and practice.

Table 2.7 Moped insurance: relativities from a multiplicative Poisson GLM for claim frequency and a gamma GLM for claim severity

Rating factor	Class	Duration	No. claims	Relativities, frequency	Relativities, severity	Relativities, pure premium
Vehicle class	1	9833	391	1.00	1.00	1.00
	2	8824	395	0.78	0.55	0.42
Vehicle age	1	1918	141	1.55	1.79	2.78
	2	16740	645	1.00	1.00	1.00
Zone	1	1451	206	7.10	1.21	8.62
	2	2486	209	4.17	1.07	4.48
	3	2889	132	2.23	1.07	2.38
	4	10069	207	1.00	1.00	1.00
	5	246	6	1.20	1.21	1.46
	6	1369	23	0.79	0.98	0.78
	7	147	3	1.00	1.20	1.20

2.4 Case Study: Motorcycle Insurance

Under the headline "case study" we will present larger exercises using authentic insurance data; for solving the case studies, a computer equipped with SAS or other suitable software is required.

The data for this case study comes from the former Swedish insurance company *Wasa*, and concerns partial casco insurance, for *motorcycles* this time. It contains aggregated data on all insurance policies and claims during 1994–1998; the reason for using this rather old data set is confidentiality—more recent data for ongoing business can not be disclosed. The data set `mccase.txt`, available at www.math.su.se/GLMbook, contains the following variables (with Swedish acronyms):

- *AGARALD*: The owners age, between 0 and 99.
- *KON*: The owners gender, M (male) or K (female).
- *ZON*: Geographic zone numbered from 1 to 7, in a standard classification of all Swedish parishes. The zones are the same as in the moped Example 1.1.
- *MCKLASS*: MC class, a classification by the so called EV ratio, defined as (Engine power in kW $\times$ 100) / (Vehicle weight in kg $+$ 75), rounded to the nearest lower integer. The 75 kg represent the average driver weight. The EV ratios are divided into seven classes as seen in Table 2.8.
- *FORDALD*: Vehicle age, between 0 and 99.
- *BONUSKL*: Bonus class, taking values from 1 to 7. A new driver starts with bonus class 1; for each claim-free year the bonus class is increased by 1. After the first claim the bonus is decreased by 2; the driver can not return to class 7 with less than 6 consecutive claim free years.

Table 2.8 Motorcycle insurance: rating factors and relativities in current tariff

Rating factor	Class	Class description	Relativity
Geographic zone	1	Central and semi-central parts of Sweden's three largest cities	7.678
	2	Suburbs plus middle-sized cities	4.227
	3	Lesser towns, except those in 5 or 7	1.336
	4	Small towns and countryside, except 5–7	1.000
	5	Northern towns	1.734
	6	Northern countryside	1.402
	7	Gotland (Sweden's largest island)	1.402
MC class	1	EV ratio –5	0.625
	2	EV ratio 6–8	0.769
	3	EV ratio 9–12	1.000
	4	EV ratio 13–15	1.406
	5	EV ratio 16–19	1.875
	6	EV ratio 20–24	4.062
	7	EV ratio 25–	6.873
Vehicle age	1	0–1 years	2.000
	2	2–4 years	1.200
	3	5– years	1.000
Bonus class	1	1–2	1.250
	2	3–4	1.125
	3	5–7	1.000

- *DURATION*: the number of policy years.
- *ANTSKAD*: the number of claims.
- *SKADKOST*: the claim cost.

The "current" tariff, the actual tariff from 1995, is based on just four rating factors, described in Table 2.8, where their current relativities are also given.

For each rating factor, we chose the class with the highest duration as base class. For the interested reader, we mention that the annual premium in the base cell $(4, 3, 3, 3)$ is 183 SEK, while the highest premium is 24 156 SEK (!).

Problem 1: Aggregate the data to the cells of the current tariff. Compute the empirical claim frequency and severity at this level.

Problem 2: Determine how the duration and number of claims is distributed for each of the rating factor classes, as an indication of the accuracy of the statistical analysis.

Problem 3: Determine the relativities for claim frequency and severity separately, by using GLMs; use the result to get relativities for the pure premium.

Problem 4: Compare the results in 3 to the current tariff. Is there a need to change the tariff? Which new conclusions can be drawn from the separate analysis in 3? Can we trust these estimates? With your estimates, what is the ratio between the highest pure premium and the lowest?

Exercises

2.1 (Section 2.1) Work out the details in the derivation of (2.2).

2.2 (Section 2.1) Suppose we have three rating factors, with two, three and five levels respectively. In how many ways could the tariff be written on list form? Each ordering of the cells is counted as one way of writing the tariff.

2.3 (Section 2.1) Show directly, without using the results in Sect. 2.1.3, that the normal distribution of Example 2.1 is reproductive; what is the value of $\mathrm{Var}(Y)$?

2.4 (Section 2.1) Verify that the expectation and variance of the gamma distribution in (2.4) are α/β and α/β^2, respectively.

2.5 (Section 2.1) Consider the reparameterization of the gamma distribution just before (2.5). Show that when ϕ and μ run through the first quadrant, α and β run through their parameter space, so that the same family of distributions is covered by the new parameterization.

2.6 (Section 2.1) An actuary studies the probability of customer renewal—that the customer chooses to stay with the insurance company for one more year—and how it varies between different groups. Let w_i be the number of customers in group i and X_i the number of renewals among these, so that by policy independence $X_i \sim Bin(w_i, p_i)$. Furthermore, let p_i be the probability under study, estimated by the key ratio $Y_i = X_i/w_i$.

Is the distribution of Y_i an EDM? If the answer is yes, determine the "canonical" parameter θ_i, an expression for ϕ, plus the functions $b(\cdot)$ and $c(\cdot)$. (In any case, we might call this the *relative binomial distribution*.)

2.7 (Section 2.1) The Generalized Pareto Distribution (GPD) is commonly used for large claims, especially in the area of reinsurance. The frequency function can be written, see e.g. Klugman et al. [KPW04, Appendix A.2.3.1],

$$f(y) = \frac{\gamma \alpha^{\gamma}}{(\alpha + y)^{\gamma+1}} \quad y > 0,$$

where $\alpha > 0$ and $\gamma > 0$. Is this distribution an EDM? If so, determine the canonical parameter θ_i and in addition ϕ and the function $b(\cdot)$.

2.8 (Section 2.1) Use well-known results for moment-generating functions, see e.g. Gut [Gu95, Theorem 3.3], to show that if the cumulant-generating function $\Psi(t)$ exists, then $E(Y) = \Psi'(0)$ and $\mathrm{Var}(Y) = \Psi''(0)$.

2.9 (Section 2.1) Derive the variance function of the relative binomial distribution, defined in Exercise 2.6.

2.10 (Section 2.1) Let X and Y be two independent random variables and let c be a constant. Show that the CGF has the following properties

(a) $\Psi_{cX}(t) = \Psi_X(ct)$,
(b) $\Psi_{X+Y}(t) = \Psi_X(t) + \Psi_Y(t)$.

2.11 (Section 2.1) Prove that the sum of independent gamma distributions with the same scale parameter β are again gamma distributed and determine the parameters. *Hint*: use the moment generating function.

2.12 (Section 2.1) Prove Theorem 2.2. *Hint*: Use the result in Exercise 2.10.

2.13 (Section 2.1)

(a) Derive $b'(\theta)$ in (2.12) from the $b(\theta)$ given by (2.10). Then derive $b''(\theta)$ for $p \geq 1$.
(b) Show that $h(\mu)$ in (2.13) is the inverse to $b'(\theta)$ in (2.12).
(c) Show that an EDM with the cumulant function $b(\theta)$ taken from (2.10) has variance function $v(\mu) = \mu^p$, so that it is a Tweedie model.

2.14 (Section 2.2) Use a multiplicative model in the moped example, Example 1.2, and write down the design matrix $\mathbf{X}$ for the list form that is implicit in Table 1.2. For the sake of simplicity, use just three zones, instead of seven.

2.15 (Section 2.2*) Derive the canonical link for the relative binomial distribution, defined in Exercise 2.6 and treated further in Exercise 2.9.

2.16 (Section 2.3) Derive (2.24) and differentiate ℓ with respect to each parameter to show that the ML equations are given by (1.6)–(1.8).

2.17 (Section 2.3) Derive the system of equations in (2.35) directly, in the special case with just two rating factors, by maximizing the log-likelihood. *Hint*: use (2.5) and (2.23).

Chapter 3
GLM Model Building

In the previous chapter we saw how to use GLMs to estimate relativities for rating factors and hence arrive at a possible new tariff. This may be sufficient if the task is just to update a given tariff structure, but every now and then one has to decide whether to add additional rating factors or change the levels for existing ones.

Example 3.1 (Moped insurance contd.) We concluded the analysis of moped insurance in Example 2.5 by noting that some levels of the rating factor *geographical zone* had very few observations. The question is then how many observations we need to trust the estimated relativities. While there is no definitive answer to this question, it is helpful to supplement the estimates with *confidence intervals*, which give a measure of the precision. Besides the rating factors in the tariff, we have access to the age and gender of the owner. The actuary might want to investigate whether these variables are important and hence should be included in the tariff. In this evaluation, *statistical tests* can give some guidance.

An advantage of using GLMs is that the theory provides standard statistical methods for hypothesis testing and confidence interval estimation. We begin this chapter by considering these tools in the context of non-life insurance pricing, along with methods for estimating the parameter ϕ.

3.1 Hypothesis Testing and Estimation of ϕ

Let $\ell(\hat{\mu})$ denote the log-likelihood for our data as a function of the estimated mean vector $\hat{\mu}$. If the number of non-redundant parameters, r, equals the number of observations (tariff cells) n, we can get a perfect fit by setting all $\hat{\mu}_i = y_i$. These are in fact the ML-estimates in this case, since they trivially satisfy the ML equations in (2.28), including the restriction of (2.31). (The latter fact actually calls for some reflection.) This case with $r = n$ is called the *saturated model*. While this model is trivial and of no practical interest, it is often used as a benchmark in measuring the goodness-of-fit of other models, since it has perfect fit. Such a measure is the

E. Ohlsson, B. Johansson, *Non-Life Insurance Pricing with Generalized Linear Models*, EAA Lecture Notes,
DOI 10.1007/978-3-642-10791-7_3, © Springer-Verlag Berlin Heidelberg 2010

scaled deviance D^*, which is defined as the likelihood-ratio-test (LRT) statistic of the model under consideration, against the saturated model. The LRT statistic is two times the logarithm of the likelihood-ratio, i.e.

$$D^* = D^*(\mathbf{y}, \hat{\boldsymbol{\mu}}) = 2\left[\ell(\mathbf{y}) - \ell(\hat{\boldsymbol{\mu}})\right]. \tag{3.1}$$

Here, it is tacitly assumed that the same ϕ is used in $\ell(\mathbf{y})$ and $\ell(\hat{\boldsymbol{\mu}})$. Let h denote the inverse of b' in the relation $\mu_i = b'(\theta_i)$ of Lemma 2.1, so that $\theta_i = h(\mu_i)$. Then (2.1) gives the following expression for D^*,

$$D^* = \frac{2}{\phi}\sum_i w_i\left(y_i h(y_i) - b(h(y_i)) - y_i h(\hat{\mu}_i) + b(h(\hat{\mu}_i))\right). \tag{3.2}$$

By multiplying this expression by ϕ, we get the (unscaled) deviance $D = \phi D^*$, which has the merit of not being dependent on ϕ. The expression "scaled" of course refers to the scaling by ϕ; recall from Lemma 2.1 that ϕ is a scale factor in the variance, too.

As a first example, suppose we have a GLM with normal distribution and all $w_i = 1$. We leave it as Exercise 3.1 to show that then

$$D(\mathbf{y}, \hat{\boldsymbol{\mu}}) = \sum_i \left(y_i - \hat{\mu}_i\right)^2. \tag{3.3}$$

Thus the deviance equals the sum of squares used to measure goodness-of-fit in the analysis of variance (ANOVA).

Example 3.2 (Deviance of the Poisson distribution) Let us return to the relative Poisson distribution in Sect. 2.1.1. The log-likelihood is found by taking logarithms in (2.3) and summing over i. From this, we get the deviance

$$\begin{aligned} D(\mathbf{y}, \hat{\boldsymbol{\mu}}) &= 2\sum_i w_i[y_i \log(y_i) - y_i] - 2\sum_i w_i[y_i \log(\hat{\mu}_i) - \hat{\mu}_i] \\ &= 2\sum_i \left[w_i y_i \log(y_i/\hat{\mu}_i) + w_i(\hat{\mu}_i - y_i)\right]. \end{aligned}$$

Note that in the Poisson case $\phi = 1$, so that $D^* = D$.

The gamma case is treated in Exercise 3.2. In summary, for the normal, Poisson and gamma distributions we have the unscaled deviances $D(\mathbf{y}, \hat{\boldsymbol{\mu}})$, where, respectively,

$$\begin{aligned} D(\mathbf{y}, \hat{\boldsymbol{\mu}}) &= \sum_i w_i(y_i - \hat{\mu}_i)^2, \\ D(\mathbf{y}, \hat{\boldsymbol{\mu}}) &= 2\sum_i w_i(y_i \log y_i - y_i \log \hat{\mu}_i - y_i + \hat{\mu}_i), \\ D(\mathbf{y}, \hat{\boldsymbol{\mu}}) &= 2\sum_i w_i(y_i/\hat{\mu}_i - 1 - \log(y_i/\hat{\mu}_i)). \end{aligned} \tag{3.4}$$

It is interesting to note that deviances may be interpreted as weighted sums of distances of the estimated means $\hat{\mu}_i$ from the observations y_i. If we define

$$d(y,\mu) = 2[yh(y) - b(h(y)) - yh(\mu) + b(h(\mu))], \tag{3.5}$$

then the deviance may be written

$$D = \sum_i w_i d(y_i, \mu_i).$$

Considered as a function of μ, for fixed y, $d(y,\mu)$ is decreasing for $\mu < y$ and increasing for $\mu > y$, and $d(y,y) = 0$. To see this, note that since h is the inverse of b',

$$\frac{\partial}{\partial \mu} d(y,\mu) = 2[-yh'(\mu) + b'(h(\mu))h'(\mu)] = 2h'(\mu)(\mu - y). \tag{3.6}$$

Thus $d(y,\mu)$ is a distance in the above sense, since by assumption b'', and thus h', is positive.

Figure 3.1 shows graphs of the function $d(1,\mu)$ for the normal, Poisson and gamma cases, i.e.

$$\begin{aligned} d(y,\mu) &= (y-\mu)^2, \\ d(y,\mu) &= 2(y\log y - y\log\mu - y + \mu), \\ d(y,\mu) &= 2(y/\mu - 1 - \log(y/\mu)). \end{aligned} \tag{3.7}$$

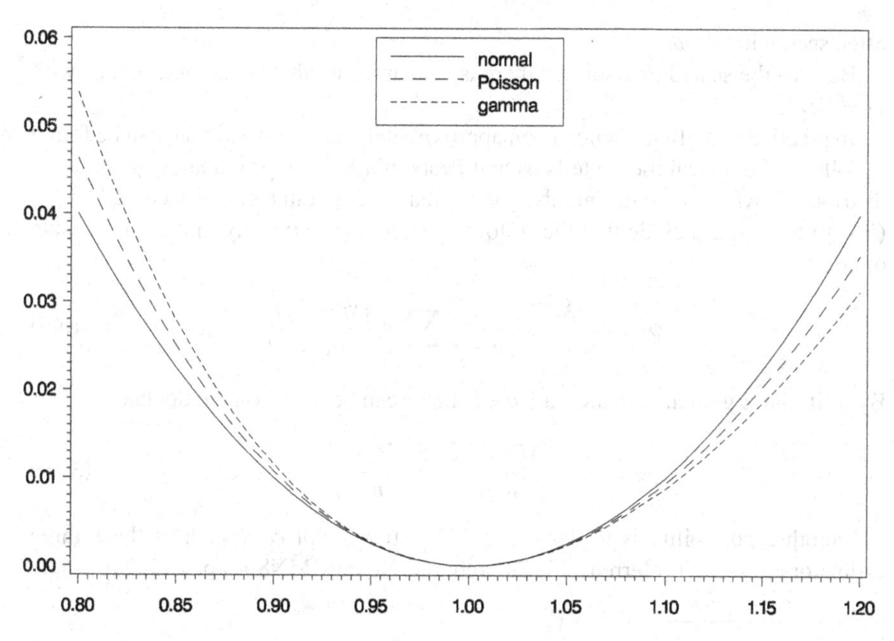

Fig. 3.1 Distance functions $d(1,\mu)$

Note that considering the variance parameter ϕ as fixed, maximizing the likelihood is trivially equivalent to minimizing the unscaled deviance. In the normal linear model, maximum likelihood is equivalent to minimizing a sum of squares. For the Poisson and gamma GLMs, we see that the deviance plays a similar role, and the figure indicates that the difference to a squared distance is not that large, if μ is close to y. This suggests that with large amounts of data, it may make little difference which deviance you use, see e.g. the examples for the multiplicative normal, Poisson and gamma models in [IJ09]. See further in Exercises 3.4 and 3.5.

3.1.1 Pearson's Chi-Square and the Estimation of ϕ

A classic measure of the goodness-of-fit of a statistical model is Pearson's chi-square X^2. In a generalized form for GLMs, this statistic is defined as[1]

$$X^2 = \sum_i \frac{(y_i - \hat{\mu}_i)^2}{\mathrm{Var}(Y_i)} = \frac{1}{\phi} \sum_i w_i \frac{(y_i - \hat{\mu}_i)^2}{v(\hat{\mu}_i)}. \tag{3.8}$$

Note from (3.3) that in the normal distribution case we have $X^2 = D^*$. In the Poisson case we have $\phi = 1$ and $v(\mu) = \mu$, and we recognize the classic Pearson chi-square statistic

$$X^2 = \sum_i w_i \frac{(y_i - \hat{\mu}_i)^2}{\hat{\mu}_i},$$

often seen with all $w_i = 1$.

Besides the scaled Pearson X^2 there is also an "unscaled" version, ϕX^2, as with $D = \phi D^*$.

In practice ϕ is often unknown; an approximately unbiased estimate can be found as follows. Statistical theory tells us that Pearson's X^2 is approximately $\chi^2(n-r)$-distributed, where r is the number of estimated β parameters. Hence, $E(X^2) \approx (n-r)$ and we conclude that the following is an approximately unbiased estimate of ϕ

$$\hat{\phi}_X = \frac{\phi X^2}{n-r} = \frac{1}{n-r} \sum_i w_i \frac{(y_i - \hat{\mu}_i)^2}{v(\hat{\mu}_i)}. \tag{3.9}$$

By a similar argument, an alternative estimator can be based on the deviance,

$$\hat{\phi}_D = \frac{\phi D^*(\mathbf{y}, \hat{\boldsymbol{\mu}})}{n-r} = \frac{D(\mathbf{y}, \hat{\boldsymbol{\mu}})}{n-r}. \tag{3.10}$$

Yet another possibility is to derive the ML estimator of ϕ. Which of these three estimators is to be preferred? McCullagh and Nelder [MN89, pp. 295–296], note

[1]The X here is actually the upper-case form of the Greek letter χ, and not an upper-case x.

for the gamma case that both the MLE and $\hat{\phi}_D$ are sensitive to rounding errors and model error (deviance from the chosen model), and recommend using $\hat{\phi}_X$, by them called the moment estimator. Meng [Me04] shows numerically that the choice of estimator can give quite different results in the gamma case and again that $\hat{\phi}_X$ is more robust against model error.

An observed claim severity y_i in cell i is in effect the mean of individual losses y_{ik}, viz. $y_i = \sum_k y_{ik}/w_i$. We say that y_i are *aggregated* data. When analyzing the claim severity, it is preferable to do the GLM analysis directly on the non-aggregated data y_{ik}. In principle, this should make no difference to the estimates $\hat{\beta}$, but when estimating ϕ we obviously loose information by aggregating.

Our conclusion is that estimation of ϕ should preferably be made using the Pearson $\hat{\phi}_X$ on non-aggregated data. In some cases the latter may not be practical, however; then $\hat{\phi}_X$ can be computed from aggregated data.

3.1.2 Testing Hierarchical Models

As an aid in deciding whether or not to include a rating factor in a GLM model we may perform a hypothesis test. This can be done with a likelihood-ratio-test (LRT) of the two models against each other, with and without the particular rating factor. The models must be hierarchically arranged with a null hypothesis that specifies a submodel of the basic model, by setting some parameters in the basic model to zero. The following result on LRTs is useful.

Lemma 3.1 (D-subtraction) *Consider two models H_r and H_s, such that $H_s \subset H_r$. Let $\hat{\boldsymbol{\mu}}^{(r)}$ be the MLEs under H_r and similarly for H_s. Then the LRT statistic for testing H_s against H_r is $D^*(\mathbf{y}, \hat{\boldsymbol{\mu}}^{(s)}) - D^*(\mathbf{y}, \hat{\boldsymbol{\mu}}^{(r)})$.*

The result is an immediate consequence of the definition (3.1). Note that in order for the models to be hierarchically ordered, they must come from the same EDM family (Poisson, gamma, ...) with a common ϕ. Under general conditions, LRTs are approximately χ^2-distributed, see e.g. [Li93, Theorem 9.1]. The degrees of freedom are $f_r - f_s$, if the models have f_r and f_s non-redundant β-parameters, respectively.

Example 3.3 (The Poisson case) For the relative Poisson distribution, the goodness-of-fit statistic was found in Example 3.2. By inserting it into Lemma 3.1, we get

$$D^*(\mathbf{y}, \hat{\boldsymbol{\mu}}^{(s)}) - D^*(\mathbf{y}, \hat{\boldsymbol{\mu}}^{(r)}) = 2\sum_i w_i y_i \log\left(\frac{\hat{\mu}_i^{(r)}}{\hat{\mu}_i^{(s)}}\right) + \sum_i w_i \left(\hat{\mu}_i^{(s)} - \hat{\mu}_i^{(r)}\right).$$

The gamma case is treated in Exercise 3.6.

If we estimate ϕ by $\hat{\phi}_X$, the LRT is modified to

$$\frac{\phi}{\hat{\phi}_X}(D^*(\mathbf{y}, \hat{\boldsymbol{\mu}}^{(s)}) - D^*(\mathbf{y}, \hat{\boldsymbol{\mu}}^{(r)})) = \frac{D(\mathbf{y}, \hat{\boldsymbol{\mu}}^{(s)}) - D(\mathbf{y}, \hat{\boldsymbol{\mu}}^{(r)})}{\hat{\phi}_X}.$$

Table 3.1 Moped insurance: test of excluding one variable at a time, multiplicative gamma model

Rating factor	$f_r - f_s$	Test statistic	p-value
Vehicle class	1	122.71	<0.0001
Vehicle age	1	79.91	<0.0001
Zone	6	7.79	0.2539

A question is under which model $\hat{\phi}_X$ is to be computed? The only reasonable choice is the larger model; in the smaller model, the estimate would include the variation that is explained by the extra variable(s), which could not be considered random if the null hypothesis would be rejected.

Example 3.4 (Moped insurance contd.) In Example 2.5, Table 2.7, the geographic zones seemed to have little effect on claim severity. It is interesting to test whether there is any significant contribution of having this rating factor in the model at all. We can use D-subtraction to perform this *effect test*—testing exclusion of one variable at a time. The result is given in Table 3.1.

While the first two rating factors are clearly significant, geographic zone shows no significant influence on claim severity. This indicates that the zone may be excluded from the analysis of claim severity. In the same test for claim frequency—not shown here—all rating factors are significant.

3.2 Confidence Intervals Based on Fisher Information

Confidence intervals is an important tool for judging the precision of estimated relativities. They provide information on whether to merge some levels of a rating factor, or to judge if a suggested change in an existing tariff is really motivated. Changes in the tariff should not be made due to random fluctuations in the estimates; a new tariff should reflect changes in the underlying expected cost only.

3.2.1 Fisher Information

To construct confidence intervals, we will need the *Fisher information* **I**, which is defined as minus the expectation of the matrix of second derivatives of the log-likelihood function ℓ. To determine **I** we differentiate $\partial\ell/\partial\beta_j$ in (2.27), this time with respect to β_k. Recall from Sect. 2.3.2 that $\partial\mu_i/\partial\eta_i = 1/g'(\mu_i)$, and $\partial\eta_i/\partial\beta_j = x_{ij}$. For $j = 1, 2, \ldots, r$; $k = 1, 2, \ldots, r$; the chain rule now yields

$$\frac{\partial^2\ell}{\partial\beta_j\partial\beta_k} = \sum_i \frac{w_i}{\phi}\frac{\partial}{\partial\mu_i}\left[\frac{y_i - \mu_i}{v(\mu_i)g'(\mu_i)}\right] x_{ij}\frac{\partial\mu_i}{\partial\eta_i}\frac{\partial\eta_i}{\partial\beta_k}$$

$$= \sum_i \frac{w_i}{\phi} \frac{\partial}{\partial \mu_i} \left[\frac{y_i - \mu_i}{v(\mu_i) g'(\mu_i)} \right] x_{ij} \frac{1}{g'(\mu_i)} x_{ik}$$

$$= -\sum_i x_{ij} a_i x_{ik}, \tag{3.11}$$

where

$$a_i = \frac{w_i}{\phi v(\mu_i) g'(\mu_i)^2} \left(1 + (y_i - \mu_i) \frac{[v(\mu_i) g''(\mu_i) + v'(\mu_i) g'(\mu_i)]}{v(\mu_i) g'(\mu_i)} \right), \tag{3.12}$$

for $i = 1, 2, \ldots, n$. The details of this last derivation are left as an exercise. By letting $\mathbf{H}$ be the matrix of second derivatives of ℓ, and forming the diagonal matrix $\mathbf{A} = \text{diag}(a_i)$, (3.11) can be written on matrix form as $\mathbf{H} = -\mathbf{X}'\mathbf{A}\mathbf{X}$. As before, $\mathbf{X}$ is the design matrix, first encountered in Sect. 2.2. Since $E(Y_i) = \mu_i$, the expected value of $\mathbf{A}$ is a diagonal matrix $\mathbf{D}$ with diagonal elements d_i equal to the first factor in a_i,

$$d_i = \frac{w_i}{\phi v(\mu_i) g'(\mu_i)^2}. \tag{3.13}$$

Thus, the Fisher information is given by

$$\mathbf{I} = -E[\mathbf{H}] = \mathbf{X}' E[\mathbf{A}] \mathbf{X} = \mathbf{X}' \mathbf{D} \mathbf{X}. \tag{3.14}$$

For our most important models, the Tweedie models of Sect. 2.1.4 with log link, we have

$$d_i = \frac{w_i}{\phi \mu_i^{p-2}}. \tag{3.15}$$

It is worth noting that the information grows linearly in w_i.

For confidence interval construction, we need an estimate of $\mathbf{I}$; such an estimate can be found by the plug-in principle, i.e., we insert our previous estimates $\hat{\mu}_i$ and $\hat{\phi}_X$ in the above expressions for d_i.

3.2.2 Confidence Intervals

Confidence intervals of GLM parameter estimates can be derived from general ML estimation theory, according to which MLEs are, under general conditions, asymptotically normally distributed and unbiased, with covariance matrix equal to the inverse of the Fisher information $\mathbf{I}$. This asymptotic result motivates the approximation

$$\hat{\boldsymbol{\beta}} \stackrel{d}{\approx} N(\boldsymbol{\beta}; \mathbf{I}^{-1}), \tag{3.16}$$

where $\stackrel{d}{\approx}$ denotes approximate distribution. In Sect. 3.2.5, we give a motivation for (3.16), even though the details of the asymptotic theory are beyond the scope of this text. Note that the covariance matrix is proportional to ϕ, since $\mathbf{I}$ is inversely so.

Table 3.2 Moped insurance: relativities for claim frequency with 95% confidence intervals, multiplicative Poisson model

Rating factor	Class	Duration	Estimated relativity	Lower limit	Upper limit
Vehicle class	1	9833	1.00	–	–
	2	8824	0.78	0.65	0.93
Vehicle age	1	1918	1.55	1.23	1.96
	2	16740	1.00	–	–
Zone	1	1451	7.10	5.52	9.13
	2	2486	4.17	3.26	5.34
	3	2889	2.23	1.69	2.94
	4	10069	1.00	–	–
	5	246	1.20	0.43	3.36
	6	1369	0.79	0.46	1.37
	7	147	1.00	0.24	4.23

By the aid of (3.16), standard normal distribution confidence intervals for β_j may be computed. More often, we are interested in the relativities $\gamma_j = \exp\{\beta_j\}$, and confidence intervals for these may be derived as follows.

- Let $\{c_{j,k}\}$ denote the elements of the covariance matrix $\mathbf{C} = \mathbf{I}^{-1}$ in (3.16). An approximate 95% confidence interval (a, b) for β_j is then $\hat{\beta}_j \pm 1.96\sqrt{c_{jj}}$.
- An approximate 95% confidence interval for γ_j is given by $(\exp(a), \exp(b))$. Note that these limits are not symmetric around the estimate $\hat{\gamma}_j = \exp\{\hat{\beta}_j\}$.

Confidence intervals for the estimated cell means, μ_i, are determined as follows.

- Construct a confidence interval (a, b) for $\eta_i = \sum_j x_{ij}\beta_j$ in the usual way, using (3.16) and noting that the $\hat{\beta}_j$'s are not independent, so that covariances must be taken into consideration.
- Use the relation $\log(\mu_i) = \eta_i$ and obtain a confidence interval for μ_i as $(\exp(a), \exp(b))$.

Example 3.5 (Moped insurance contd.) In Example 2.5 we presented estimates of the relativities γ_j. We are now able to supplement these estimates with confidence intervals. In the case of claim frequency, the results are given in Table 3.2.

A striking feature of these intervals is that they are quite broad, even for levels with several thousands of observations. The data shows no evidence for separate pricing of the geographic areas 5–7, as compared to the other rural areas in 4. Of course, this does not necessarily mean that there are no underlying differences, but without enough data, it is questionable to separate 5–7 from 4. All other parameters are found to be of significant importance for the claim frequency. Note the importance of choosing a base level with large exposure—with zone 7 as base all the zone parameters become very uncertain and it would be hard to draw any conclusions.

Table 3.3 Moped insurance: relativities for claim severity with 95% confidence intervals, multiplicative gamma model

Rating factor	Class	No. claims	Estimated relativity	Lower limit	Upper limit
Vehicle class	1	391	1.00	–	–
	2	395	0.55	0.49	0.61
Vehicle age	1	141	1.79	1.57	2.05
	2	645	1.00	–	–
Zone	1	206	1.21	1.05	1.41
	2	209	1.07	0.93	1.24
	3	132	1.07	0.91	1.25
	4	207	1.00	–	–
	5	6	1.21	0.67	2.18
	6	23	0.98	0.72	1.34
	7	3	1.20	0.53	2.73

In the same way, we get confidence intervals for claim severity, displayed in Table 3.3. These results confirm those of Example 3.4 that the geographic zone is of no significant importance, with a possible exception for zone 1. An additional test, not presented here, where zones 2–7 were merged into one, showed no significance of this rating factor. In practice, we would combine all this quantitative information with qualitative "know-how" of a moped insurance expert before drawing our final conclusions.

Our next issue is the construction of confidence intervals for the *pure premium* relativities. These are of interest when we are in the process of changing an existing tariff. Let γ^F and γ^S be the relativity for claim frequency and severity, respectively, for some level of a rating factor; here we omit the index j for simplicity in notation. Then the pure premium has relativity $\gamma^P = \gamma^F \gamma^S$. This way to multiply estimated relativities was used already in Table 2.7.

Again, we use the log scale when constructing confidence intervals, with $\beta^F = \log(\gamma^F)$, $\beta^S = \log(\gamma^S)$ and $\beta^P = \log(\gamma^P)$, so that $\hat{\beta}^P = \hat{\beta}^F + \hat{\beta}^S$. Now, if the frequency and severity were independent, we could simply add the variances from the separate analysis of claim frequency and severity to get $\text{Var}(\hat{\beta}^P)$. However, the fact that the severity depends on the number of claims indicates that there could be some dependence. A further complication is that the analysis of the severity is made conditionally on the number of claims $\mathbf{N} = N_1, \ldots, N_n$. Note again that the observed values of N_i are called w_i in that analysis.

It is seen from (3.16) that the GLM estimates are approximately unbiased, which for claim severity means that $E(\hat{\beta}^S|\mathbf{N}) \approx \beta^S$. By using this and the fact that $\hat{\beta}^F$ is based on $\mathbf{N}$ only, so that $\text{Var}(\hat{\beta}^F|\mathbf{N}) = 0$, we get

$$\text{Var}(\hat{\beta}^P) = \text{Var}[E(\hat{\beta}^F + \hat{\beta}^S|\mathbf{N})] + E[\text{Var}(\hat{\beta}^F + \hat{\beta}^S|\mathbf{N})] \approx \text{Var}[\hat{\beta}^F] + E[\text{Var}(\hat{\beta}^S|\mathbf{N})].$$

Table 3.4 Moped insurance: relativities for the pure premium, with 95% confidence intervals based on (3.17)

Rating factor	Class	Duration	No. claims	Tariff relativity	Estimated relativity	Lower limit	Upper limit
Vehicle class	1	9833	391	1.00	1.00	–	–
	2	8825	395	0.50	0.42	0.34	0.52
Vehicle age	1	1918	141	1.66	2.78	2.12	3.64
	2	16740	645	1.00	1.00	–	–
Zone	1	1451	206	5.16	8.62	6.44	11.53
	2	2486	209	3.10	4.48	3.37	5.96
	3	2889	132	1.92	2.38	1.73	3.27
	4	10069	207	1.00	1.00	–	–
	5	246	6	2.50	1.46	0.45	4.75
	6	1369	23	1.50	0.78	0.41	1.46
	7	147	3	1.00	1.20	0.23	6.31

From the GLM analysis of claim frequency we get $\mathrm{Var}[\hat{\beta}^F]$, but for claim severity we actually get the *conditional* variance $\widehat{\mathrm{Var}}(\hat{\beta}^S|\mathbf{N})$, which is trivially an unbiased estimate of $E[\mathrm{Var}(\hat{\beta}^S|\mathbf{N})]$. Hence an estimate of the variance of $\hat{\beta}^P$ is found as

$$\widehat{\mathrm{Var}}(\hat{\beta}^P) = \widehat{\mathrm{Var}}(\hat{\beta}^F) + \widehat{\mathrm{Var}}(\hat{\beta}^S|\mathbf{N}). \tag{3.17}$$

Note that this is just the simple adding of variances that we should have used, had the frequency and the severity been independent. So the method is very simple, it is just the motivation that is somewhat complicated. We can use the resulting $\mathrm{Var}(\hat{\beta}^P)$ to compute an approximative confidence interval for the pure premium, with the same method as for claim frequency or severity above.

Remark 3.1 A more elegant way of motivating the use of (3.17) would be to claim that $\mathrm{Var}[\hat{\beta}^F] + \mathrm{Var}(\hat{\beta}^S|\mathbf{N})$ is more *relevant* as a measure of the quadratic error in $\hat{\beta}^P$, than $\mathrm{Var}[\hat{\beta}^F] + E[\mathrm{Var}(\hat{\beta}^S|\mathbf{N})]$; this is in line with Corollary 2 in Sundberg [Su03], to which we refer for further discussion of this issue.

Example 3.6 (Moped insurance contd.) Approximate 95% confidence intervals for the pure premium in the moped example are given in Table 3.4. While our previous analysis indicated that the relativity for *vehicle class* 2 could be decreased from 0.50 to 0.42, this is no longer obvious since 0.50 lies within the confidence interval. We need additional information to decide this finally, e.g. whether 0.50 was based on a calculation or only merely on judgement. It is clear, however, that *vehicle age* 1 should have a higher relativity, and the same goes for the three first zones—while the relativity of zone 3 is within the interval, it is reasonable not to increase the difference to classes 1 and 2 just because of that; we must always remember that it

Table 3.5 Moped insurance: relativities for risk premium. Poisson model with merged zones, plus Gamma model without zones

Rating factor	Class	Duration	No. claims	Estimated relativity	Current relativity
Vehicle class	1	9833	391	1.00	1.00
	2	8825	395	0.45	0.50
Vehicle age	1	1918	141	2.92	1.66
	2	16740	645	1.00	1.00
Zone	1	1451	206	7.24	5.16
	2	2486	209	4.26	3.10
	3	2889	132	2.28	1.92
	4	10069	207	1.00	1.00
	5	246	6	1.00	2.50
	6	1369	23	1.00	1.50
	7	147	3	1.00	1.00

is not only the relation to the base level that matters, although that is what we see in the tables.

Concerning the *zone* factor, we have seen earlier that the tariff could be somewhat simplified. For claim severity we could set all zone relativities equal to one, while for claim frequency we could do the same for zones 5–7. The latter is not the only choice though: if the northern part of Sweden had not been treated separately from the beginning, then some of the northern towns in zone 5 would actually have been classified as belonging to zone 2 or 3; without access to the data on a lower level, this problem can not be properly solved. We will return to the complicated problem of geographic classification in Chap. 4; here we can just suggest merging zones 4–7.

In the resulting tariff, all parameters that are not set to 1 are significant. It should be noted, though, that confidence and significance levels are no longer strictly valid after all these changes in classification—this is a good reason for making a follow up study when new data are available after a year or two. Our last table, Table 3.5, is a comparison of our suggested new tariff to the current tariff, after multiplying the relativities for frequency and severity. Note that the estimates for zones 1–3 are the ones for claim frequency only. The result is quite far from the tariff, so there is really need for a change, but for market reasons this change might be done gradually rather than in one step.

3.2.3 Numerical Equation Solving*

Typically, the ML equations in (2.28) do not have an explicit solution, with the notable exception of the general linear model. Instead we have to resort to some numerical method; here we shall briefly describe the widely used Newton-Raphson's method and Fisher's scoring method.

Newton-Raphson's method for the iterative solution of *a single* equation $f(x) = 0$ is well known: starting with some initial value $x^{[0]}$, each following iteration step is given by

$$x^{[n+1]} = x^{[n]} - f'(x^{[n]})^{-1} f(x^{[n]}), \tag{3.18}$$

where the super-index $[n]$ denotes iteration number. The method has a straight-forward generalization to the case with r equations and r unknown variables, with the matrix of first derivatives replacing f'. Recall the matrix formulation of the ML equations in (2.32); in our case each iteration step of Newton-Raphson's method is given by

$$\boldsymbol{\beta}^{[n+1]} = \boldsymbol{\beta}^{[n]} - (\mathbf{H}^{-1})^{[n]} \left(\mathbf{X}'\mathbf{W}^{[n]}\mathbf{y} - \mathbf{X}'\mathbf{W}^{[n]}\boldsymbol{\mu}^{[n]}\right), \tag{3.19}$$

where $\mathbf{H}$ was defined in Sect. 3.2.1. Note that $\mathbf{H}$, $\mathbf{W}$ and $\boldsymbol{\mu}$ must be recalculated in each step.

Fisher's scoring-method replaces $\mathbf{H}$ by its expected value, the negative of the Fisher information $-\mathbf{I}$.

$$\boldsymbol{\beta}^{[n+1]} = \boldsymbol{\beta}^{[n]} + (\mathbf{I}^{-1})^{[n]} \left(\mathbf{X}'\mathbf{W}^{[n]}\mathbf{y} - \mathbf{X}'\mathbf{W}^{[n]}\boldsymbol{\mu}^{[n]}\right). \tag{3.20}$$

In the special case of canonical link, discussed in Sect. 2.2.1, the methods are identical, cf. Exercise 3.10. Otherwise we have to choose one of them, as will now be discussed.

Fisher's method is somewhat easier to implement, a fact that is seldom of importance because of the availability of statistical software. As opposed to $\mathbf{H}$, the matrix $\mathbf{I}$ is always positive definite (except in pathological cases); therefore one may expect more stable convergence to the target by using Fisher's rather than Newton-Raphson's method. On the other hand, the latter method is faster and when we are close to the target $\mathbf{H}$ is also positive definite. This suggests that we could start with a few steps of Fisher's scoring method and then change to Newton-Raphson when we approach the target.

*3.2.4 Do the ML Equations Really Give a Maximum?**

By differentiating the log-likelihood and setting the result equal to zero, we derived the ML equations in (2.28), whose solution provides us with a stationary point of the log-likelihood-function. But is it really a maximum? A full exploration of this question is beyond the scope of this text, but we will try to give part of the answer by narrowing the focus to the case with canonical link in Sect. 2.2.1. We use the notation of Sect. 3.2.1.

By Exercise 3.10, $\mathbf{A} = \mathbf{D} = \text{diag}(w_i v(\mu_i)/\phi)$, from which it follows that $\mathbf{A}$ is positive definite. By choosing our rating factors appropriately, we can avoid that $\mathbf{X}$

contains linearly dependent columns. Then $\mathbf{H} = -\mathbf{X}'\mathbf{D}\mathbf{X}$ is negative definite and the log-likelihood function ℓ must be concave. The conclusion is that the ML equations give a unique maximum.

Even in the more complicated case of an arbitrary link function, a maximum, if it does exist, is given by a solution to the ML equations, but there is a theoretical possibility that these equations have more than one solution.

3.2.5 Asymptotic Normality of the ML Estimators*

In (3.16), it was stated that the ML estimator $\hat{\boldsymbol{\beta}} = (\hat{\beta}_1, \ldots, \hat{\beta}_r)$ has an approximate normal distribution. A rigorous asymptotic result to support this statement is beyond the scope of this text, but in this section we give a motivation for why (3.16) may be expected.

The result needed from probability theory is a multivariate central limit theorem. The theorem concerns random vectors $\mathbf{S}_n = (S_{n1}, \ldots, S_{nr})$, where $S_{nj} = \sum_{i=1}^{n} X_{ij}$. The vectors $\mathbf{X}_i = (X_{i1}, \ldots, X_{ir})$; $i = 1, 2, \ldots$; are assumed independent. However, it is not assumed that the components $X_{i1}, \ldots, X_{ir}$ of each vector are independent. We assume that $E[X_{ij}] = 0$ for each i, j (if this is not the case, it is easily achieved by subtracting the means of each variable). We denote by $\mathbf{V}_n$ the matrix with elements $v_{jk} = \mathrm{Cov}(S_{nj}, S_{nk}) = \sum_{i=1}^{n} \mathrm{Cov}(X_{ij}, X_{ik})$. Covariance matrices like $\mathbf{V}_n$ are symmetric non-negative definite, and such matrices have a *root*: there exists a matrix denoted $\mathbf{V}_n^{\frac{1}{2}}$, such that $\mathbf{V}_n = \mathbf{V}_n^{\frac{1}{2}} \mathbf{V}_n^{\frac{1}{2}}$. If $\mathbf{V}_n$ is positive definite, this root matrix also has an inverse $\mathbf{V}_n^{-\frac{1}{2}}$. Let $N(\mathbf{0}, \mathbb{I})$ denote the multivariate normal distribution with a mean vector consisting of zeros and a covariance matrix which is the identity matrix $\mathbb{I}$. Under certain conditions on its asymptotic behavior, the multivariate central limit theorem states that

$$\mathbf{V}_n^{-\frac{1}{2}} \mathbf{S}_n \overset{d}{\approx} N(\mathbf{0}, \mathbb{I}), \tag{3.21}$$

when the number of observations is large. Here we have interpreted $\mathbf{S}_n$ as a column vector.

To see how this central limit theorem comes into play in our case, define the functions $s_j(\mathbf{y}, \boldsymbol{\beta})$ by

$$s_j(\mathbf{y}, \boldsymbol{\beta}) = \sum_{i=1}^{n} \frac{w_i}{\phi} \frac{y_i - \mu_i}{v(\mu_i) g'(\mu_i)} x_{ij}; \quad j = 1, \ldots, r, \tag{3.22}$$

where n is the number of observations. From (2.28) we see that the ML estimator $\hat{\boldsymbol{\beta}} = (\hat{\beta}_1, \ldots, \hat{\beta}_r)$ is obtained as the solution to the system of equations

$$s_j(\mathbf{y}, \boldsymbol{\beta}) = 0; \quad j = 1, \ldots, r. \tag{3.23}$$

Let $\boldsymbol{\beta}^0 = (\beta_1^0, \ldots, \beta_r^0)$ denote the 'true' value of $\boldsymbol{\beta}$. Since the Y_i's are assumed independent, it follows from (3.22) that each $s_j(\mathbf{Y}, \boldsymbol{\beta}^0)$ is a sum of independent random variables with mean 0. Furthermore, it is readily verified (see Exercise 3.11) that the matrix $\mathbf{V}_n$ with elements $v_{jk} = \mathrm{Cov}(s_j(\mathbf{Y}, \boldsymbol{\beta}^0), s_k(\mathbf{Y}, \boldsymbol{\beta}^0))$ is equal to the Fisher information matrix $\mathbf{I}$ in (3.14). Thus by the central limit theorem in (3.21),

$$\mathbf{I}^{-\frac{1}{2}}\mathbf{s}(\mathbf{y}, \boldsymbol{\beta}) \overset{d}{\approx} N(\mathbf{0}, \mathbb{I}), \tag{3.24}$$

where $\mathbf{s}(\mathbf{y}, \boldsymbol{\beta})$ is the column vector with elements $s_j(\mathbf{y}, \boldsymbol{\beta})$; $j = 1, \ldots, r$.

We now consider how to go from this result to the asymptotic distribution of $\hat{\boldsymbol{\beta}} = (\hat{\beta}_1, \ldots, \hat{\beta}_r)$.

A first order approximation of $s_j(\mathbf{y}, \boldsymbol{\beta})$, considered as a function of $\boldsymbol{\beta} = (\beta_1, \ldots, \beta_r)$, around $\boldsymbol{\beta}^0 = (\beta_1^0, \ldots, \beta_r^0)$, is given by

$$s_j(\mathbf{y}, \boldsymbol{\beta}) \approx s_j(\mathbf{y}, \boldsymbol{\beta}^0) + \sum_{k=1}^{r} \left(\frac{\partial}{\partial \beta_k} s_j(\mathbf{y}, \boldsymbol{\beta}^0) \right) (\beta_k - \beta_k^0). \tag{3.25}$$

(A strict motivation of this approximation can be based on the mean value theorem of differential calculus.) From (3.11) and (3.12) it is seen that

$$\frac{\partial}{\partial \beta_k} s_j(\mathbf{y}, \boldsymbol{\beta}^0) = -\sum_{i=1}^{n} x_{ij} a_i x_{ik}, \tag{3.26}$$

where a_i is given in (3.12).

In the case of canonical link, as for Poisson with log link, it was seen in Exercise 3.10 that $a_i = d_i$, with d_i given in (3.13), so that

$$\frac{\partial}{\partial \beta_k} s_j(\mathbf{Y}, \boldsymbol{\beta}^0) = -\sum_{i=1}^{n} x_{ij} d_i x_{ik} = -I_{jk}, \tag{3.27}$$

where I_{jk} denotes the element in row j and column k of the Fisher information matrix $\mathbf{I}$. When the link is not canonical, it can be shown that the first equality in (3.27) holds approximately, when the number of observations is large. Returning to (3.25), this motivates the approximation

$$s_j(\mathbf{y}, \boldsymbol{\beta}) \approx s_j(\mathbf{y}, \boldsymbol{\beta}^0) - \sum_{k=1}^{r} I_{jk}(\beta_k - \beta_k^0). \tag{3.28}$$

Inserting $\hat{\boldsymbol{\beta}}$ for $\boldsymbol{\beta}$ in (3.28) and using the basic fact that $\hat{\boldsymbol{\beta}}$ solves (3.23), we get

$$\sum_{k=1}^{r} I_{jk}(\hat{\beta}_k - \beta_k^0) \approx s_j(\mathbf{y}, \boldsymbol{\beta}^0).$$

Writing this on matrix form, we have

$$\mathbf{I}(\hat{\boldsymbol{\beta}} - \boldsymbol{\beta}^0) \approx \mathbf{s}(\mathbf{y}, \boldsymbol{\beta}^0). \tag{3.29}$$

Thus, $\mathbf{I}^{\frac{1}{2}}(\hat{\boldsymbol{\beta}} - \boldsymbol{\beta}^0) \approx \mathbf{I}^{-\frac{1}{2}}\mathbf{s}(\mathbf{y}, \boldsymbol{\beta}^0)$ and by (3.24) the distribution of this quantity is approximately $N(\mathbf{0}, \mathbb{I})$ when the number of observations is large. Finally, (3.16) is just a rearrangement of this result, with $\boldsymbol{\beta} = \boldsymbol{\beta}^0$.

3.3 Residuals

As with ordinary linear regression, residuals can be used in GLMs to study the model fit to the data and in particular to detect outliers and check the variance assumption. Perhaps the most common choice is *Pearson residuals*

$$r_{Pi} = \frac{y_i - \hat{\mu}_i}{\sqrt{v(\hat{\mu}_i)/w_i}}. \tag{3.30}$$

The sum of these residuals is the unscaled Pearson X^2, i.e., $\sum_i r_{Pi}^2 = \phi X^2$ from Sect. 3.1.1. If μ_i were known, these Pearson residuals would have zero expectation and constant variance ϕ. *Standardized* Pearson residuals are Pearson residuals divided by $\sqrt{\phi(1-h_i)}$, where h_i is a correction for the fact that μ_i is estimated.

For the interested reader, we state without proof that h_i is an element of the so called hat matrix,

$$\mathbf{D}^{1/2}\mathbf{X}(\mathbf{X}'\mathbf{D}\mathbf{X})^{-1}\mathbf{X}'\mathbf{D}^{1/2}.$$

Alternatively, one may use the *deviance residuals*, defined as

$$r_{Di} = \sqrt{w_i d(y_i, \hat{\mu}_i)} \times \text{sign}(y_i - \hat{\mu}_i), \tag{3.31}$$

where sign is the function that returns the sign (plus or minus) of the argument, while $d(\cdot,\cdot)$ was defined in (3.5). It follows by construction that $\sum_i r_{Di}^2 = \phi D^*$. *Standardized* deviance residuals are obtained by dividing by the same factor as in the Person case. The deviance residuals are natural constructions if we recall the interpretation of $d(\mathbf{y}, \cdot)$ as a distance in Sect 3.1.

Example 3.7 (Poisson) In the case of relatively Poisson distributed claim frequencies, we have

$$r_{Pi} = \frac{y_i - \hat{\mu}_i}{\sqrt{\hat{\mu}_i/w_i}},$$

and by Example 3.2 we get

$$r_{Di} = \text{sign}(y_i - \hat{\mu}_i)\sqrt{2w_i\left[y_i \log\left(\frac{y_i}{\hat{\mu}_i}\right) - (y_i - \hat{\mu}_i)\right]}.$$

Residuals are graphed in an index plot (against observation number) or normal probability plot, to find outliers that might be erroneous or otherwise not suitable to include in the model. Another common plot is that against μ_i, in which the appropriateness of the variance function can be evaluated; if, for example, the residuals tend to "fan out" towards the right, then we might not have chosen a large enough p in a Tweedie model.

3.4 Overdispersion

In Sect. 2.1.1 we mentioned that the assumption of homogeneity within tariff cells, which lies behind the Poisson model for claim frequencies, can be questioned. Random variation between customers and insured objects, and effects of explanatory variables that are not included in the model, lead to *overdispersion*: the variance of the observations within a tariff cell is larger than the variance of a Poisson distribution.

A common way to model overdispersion is to view the mean in the Poisson distribution as a random variable. Suppose that $\Lambda_1, \Lambda_2, \ldots$ is a sequence of independent random variables with distributions on $(0, \infty)$. Furthermore, let $X_1, X_2, \ldots$ be independent random variables, such that given $\Lambda_i = \lambda_i$, X_i has a Poisson distribution with mean λ_i. By the basic properties of the Poisson distribution, $E(X_i | \Lambda_i) = \Lambda_i$ and $\mathrm{Var}(X_i | \Lambda_i) = \Lambda_i$, and so $E[X_i] = E[E(X_i | \Lambda_i)] = E[\Lambda_i]$ and the variance of X_i is

$$\mathrm{Var}[X_i] = E[\mathrm{Var}(X_i | \Lambda_i)] + \mathrm{Var}[E(X_i | \Lambda_i)] = E[\Lambda_i] + \mathrm{Var}[\Lambda_i]. \tag{3.32}$$

Assuming a mean-variance relationship for the $\{\Lambda_i\}$ induces one for the $\{X_i\}$. One possible assumption concerning the relation between $\mathrm{Var}[\Lambda_i]$ and $E[\Lambda_i]$ which turn out to have interesting implications is

$$\mathrm{Var}[\Lambda_i] = \nu E[\Lambda_i], \tag{3.33}$$

for some $\nu > 0$. If X_i is the number of claims with expectation $w_i \mu_i$, this entails, using (3.32),

$$\mathrm{Var}[X_i] = (1 + \nu) w_i \mu_i .$$

Switching to the claim frequency $Y_i = X_i / w_i$, we have $\mathrm{Var}[Y_i] = \mathrm{Var}[X_i]/w_i^2$, and so

$$\mathrm{Var}[Y_i] = (1 + \nu) \mu_i / w_i . \tag{3.34}$$

Writing $\phi = 1 + \nu$, this becomes $\mathrm{Var}[Y_i] = \phi \mu_i / w_i$. Note that this looks like the variance for the Poisson distribution with an added dispersion parameter $\phi > 1$.

There are of course many conceivable mean-variance relationships for Λ_i, for instance

$$\mathrm{Var}[\Lambda_i] = \nu (E[\Lambda_i])^2, \tag{3.35}$$

leading to

$$\mathrm{Var}[Y_i] = \mu_i / w_i + \nu \mu_i^2 . \tag{3.36}$$

To specify a distribution of X_i, one has to make a distributional assumption for Λ_i. A popular choice is the gamma distribution, since this leads to analytically tractable results. If we assume that Λ_i has a gamma distribution satisfying (3.33), one can show that the frequency function for X_i becomes (see Exercise 3.13)

$$P(X_i = x_i) = \frac{\Gamma(w_i\mu_i/\nu + x_i)}{\Gamma(w_i\mu_i/\nu)x_i!}\left(\frac{1}{1+\nu}\right)^{w_i\mu_i/\nu}\left(\frac{\nu}{1+\nu}\right)^{x_i}. \tag{3.37}$$

With (3.35), one gets (see Exercise 3.16)

$$P(X_i = x_i) = \frac{\Gamma(1/\nu + x_i)}{\Gamma(1/\nu)x_i!}\left(\frac{1}{\nu w_i\mu_i + 1}\right)^{1/\nu}\left(\frac{\nu w_i\mu_i}{\nu w_i\mu_i + 1}\right)^{x_i}. \tag{3.38}$$

Both these examples are *negative binomial* distributions with different parametrizations.

Assuming the common multiplicative model (2.31), it is straightforward to derive ML equations for $\{\beta_j\}$ and ν from (3.37) or (3.38), see Exercises 3.14 and 3.16. The parameter ν may also be estimated using the method of moments. For additional details on these models and so called *generalized Poisson regression models*, see Ismail and Jemain [IJ07].

Example 3.8 This example is based on motor third party liability data from a lorry insurance portfolio. The claim frequencies were analyzed using four rating factors: geographical region, vehicle age, vehicle model classification and annual mileage. The duration of each individual policy was one year or less. To study the phenomenon of overdispersion, the policies were randomly divided into groups so that the total duration of each group was approximately 100 policy years. The number of groups was 5 746.

We let N_i denote the number of claims in group i, and assume to begin with that the groups are completely homogeneous and disregard the rating factors. This means that the variables $N_1, N_2, \ldots$ are independent, each having a Poisson distribution with mean 100μ, where μ is the expected number of claims of a policy with duration 1. If this (unrealistic) assumption was true, a diagram of the proportion of groups having k claims, $k = 0, 1, \ldots$ would resemble a diagram of the Poisson distribution with mean 100μ. The average number of claims in the groups was 7.03 and in Fig. 3.2 the observed proportions of groups with a certain number of claims are shown together with the proportions that would be expected if the Poisson assumption was true. The observed variation between the groups is seen to be considerably larger than the expected variation.

Next, we studied what happens if we take into consideration the difference between the policies due to the rating factors. Using the estimated relativities, we calculated the expected number of claims μ_{ij} for policy j of group i, based on the values of the rating factors for the policy. Since we assume the policies to be independent, and the number of claims of each policy to have a Poisson distribution, each N_i would have a Poisson distribution with mean $\sum_j \mu_{ij}$. If $I(N_i = k)$ is the random variable which equals 1 if $N_i = k$ and 0 otherwise, the expected number

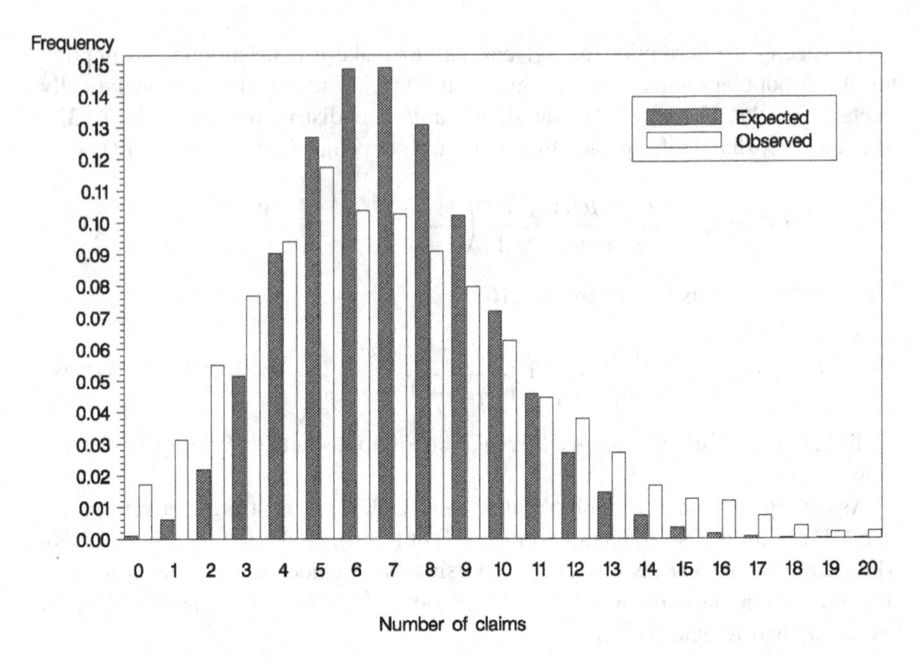

Fig. 3.2 The proportion of groups of policies with a certain number of claims for a lorry portfolio, Poisson case without rating factors

of groups with k claims is $E\sum_i I(N_i = k) = \sum_i P(N_i = k)$, and dividing this by the number of groups we get the expected proportion of groups having k claims. In Fig. 3.3 it can be seen that the fit has improved a little, but the deviation from the observed proportions is still considerable.

Of course, we could continue to include additional rating factors into the model, provided we had access to the relevant information for each policy, hoping to come closer to the observed variation in the data. However, there is variation between the policies that we can never hope to incorporate in a statistical model, most importantly varying driving behavior between the customers, but also random variation between the vehicles, roads, weather conditions and so on, which can only to a limited extent be captured by explanatory variables.

Finally, we fitted the model (3.37) to the data, retaining the rating factors. One can show (Exercise 3.13) that each N_i has a negative binomial distribution with mean $\sum_j \mu_{ij}$ and variance $(1+\nu)\sum_j \mu_{ij}$. Thus, just as in the Poisson case with rating factors, we may calculate the estimated proportion of groups having k claims and compare with the observed proportion. The result is shown in Fig. 3.4, where it is seen that the fit is now better, if still not perfect.

While overdispersion does not affect the estimates of the β-parameters, it is important to take it into consideration when it comes to constructing confidence intervals. Neglecting the additional variation yields confidence intervals that are too

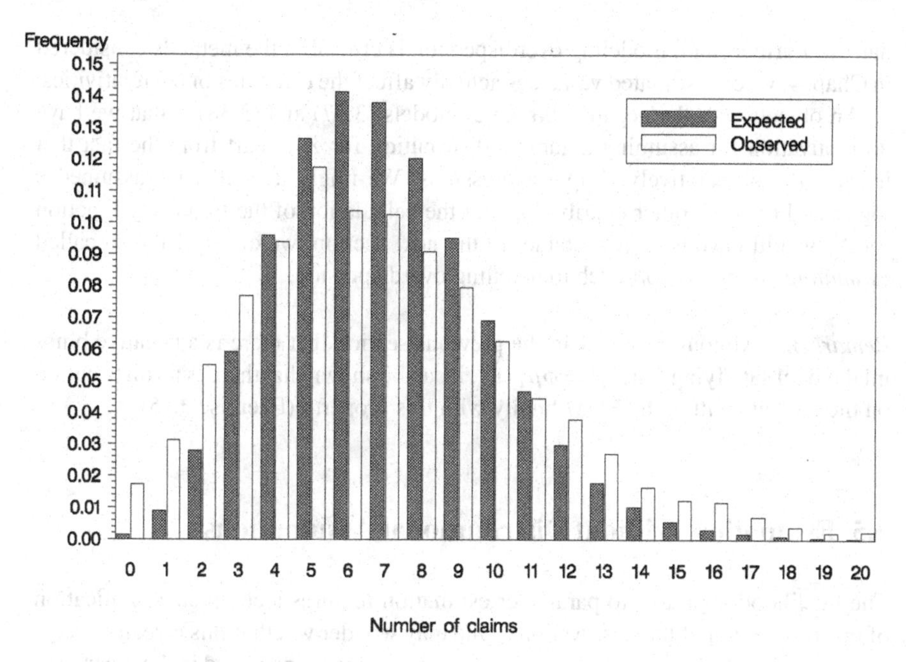

Fig. 3.3 The proportion of groups of policies with a certain number of claims for a lorry portfolio, Poisson case with rating factors

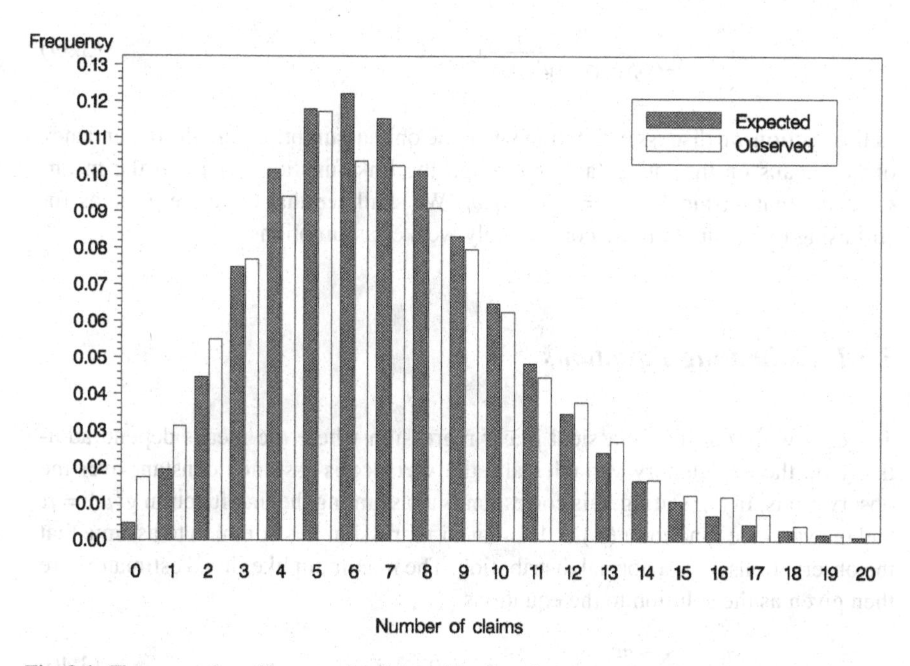

Fig. 3.4 The proportion of groups of policies with a certain expected number of claims for a lorry portfolio, negative binomial case with rating factors

narrow. Furthermore, modeling overdispersion is crucial in the methods considered in Chap. 4, where estimated variances actually affect the estimates of the relativities.

An objection to the negative binomial models (3.37) and (3.38) is that we have no motivation for assuming a gamma distribution for Λ_i, apart from the fact that it leads to comparatively simple expressions. We might as well have assumed a lognormal or some other distribution, but the calculation of the frequency function for X_i would have been intractable. In the next section, we consider the so called *estimating equations* approach to handling overdispersion.

Remark 3.2 Although we saw in the previous section that there is a negative binomial model satisfying $\text{Var}[Y_i] = \phi\mu_i/w_i$, it can be shown that there is no distribution on the integers within the EDM family with this property (Exercise 3.15).

3.5 Estimation Without Distributional Assumptions

The likelihood approach to parameter estimation requires a complete specification of the distribution of the observations. One may wonder whether this is really necessary, since in non-life insurance pricing, we are primarily interested in the means μ_i, and how the means are affected by the rating factors. In Sect. 2.3.2 we derived the equations (2.28) by the principle of maximum likelihood:

$$\sum_i \frac{w_i}{\phi} \frac{y_i - \mu_i}{v(\mu_i)g'(\mu_i)} x_{ij} = 0; \quad j = 1, \ldots, r. \tag{3.39}$$

In this section we discuss estimation when the only assumptions are the dependency of the means on the rating factors through the link function $g(\mu)$ and the mean-variance relationship $V[Y_i] = \phi v(\mu_i)/w_i$. We shall see that there are grounds for still using (3.39) under these considerably weaker assumptions.

3.5.1 Estimating Equations

To begin with, consider classical linear regression where the means depend additively on the explanatory variables and the variance is assumed constant over the observations. In our setting, this corresponds to assuming the link function $g(\mu) = \mu$ and the variance function $v(\mu) = 1$. In linear regression it is common to assume that the observations have a normal distribution. The maximum likelihood estimators are then given as the solution to the equations

$$\sum_i \frac{w_i}{\phi}(y_i - \mu_i)x_{ij} = 0; \quad j = 1, \ldots, r, \tag{3.40}$$

which in this case have an explicit solution. If $\mathbf{W}$ is the diagonal matrix with the weights w_i on the diagonal and $\mathbf{X}$ is the design matrix, then the estimators of the β-parameters may be written on matrix form as

$$\hat{\boldsymbol{\beta}} = (\mathbf{X}'\mathbf{W}\mathbf{X})^{-1}\mathbf{X}'\mathbf{W}\mathbf{y}.$$

Thus the estimators are linear functions of the observations.

The classical *least squares method* does not rely on the likelihood principle but the idea is to minimize the distance between the observations and the means, measured by a weighted sum of squares. It is well-known that the resulting estimators are the same. The choice to measure the distance between the observations and means using the quadratic function is really somewhat arbitrary, except that it is simple. However, there is a motivation for using it, given by the so called *Gauss-Markov theorem* (see Wood [Wo06, Sect. 1.3.8]). This theorem states that within the set of linear unbiased estimators, the least squares estimators have minimum variance.

To prepare for the case of general link and variance functions, we observe that linear unbiased estimators can be represented as solutions to the equations

$$\sum_i c_{ji}(y_i - \mu_i) = 0; \quad j = 1, \ldots, r, \tag{3.41}$$

where $\{c_{ij}\}$ are constants (see Exercise 3.17). Thus the Gauss-Markov Theorem says that within this family of equations, the variance is minimized by taking $c_{ij} = w_i x_{ij}$ (whether we include ϕ or not in the equations is of course irrelevant).

Now assuming arbitrary link and variance functions, consider equations of the form

$$\sum_i c_{ji}(\mu_i)(y_i - \mu_i) = 0; \quad j = 1, \ldots, r, \tag{3.42}$$

for some functions $\{c_{ji}(\cdot)\}$ of sufficient regularity. The idea of constructing estimators starting from a set of equations is called the *estimating equations* approach. Since the observations enter in a linear way on the left-hand side of (3.42), such equations are called *linear estimating equations*. As with linear estimators, the main reason for choosing them is their simplicity. Also, they are well suited for solving with the Newton-Raphson algorithm. Note that (3.39) are linear estimating equations.

There is an extension of the Gauss-Markov Theorem to the case of arbitrary $v(\mu)$ and $g(\mu)$, which roughly states that within the family of linear estimating equations (3.42) the estimators obtained as solutions to (3.39) have minimum asymptotic variance. For a more precise statement, see McCullagh [McC83]. This result gives a motivation, without reference to a particular distribution, for the way the variance function $v(\mu)$ and the link function $g(\mu)$ occur in the left hand side of the estimating equations (3.39).

The heuristic derivation of asymptotic normality in Sect. 3.2.5 depended only on the fact that the estimates satisfy these equations and the variance assumption $V[Y_i] = \phi v(\mu_i)/w_i$. This is because the central limit theorem, which lie behind the asymptotic results, relies only on the first and second order moments. Thus, the construction of confidence intervals in Sect. 3.2.2 remains valid, with $\mathbf{I}$ *defined* as $\mathbf{X}'\mathbf{D}\mathbf{X}$ and d_i as in (3.13).

Without distributional assumptions, ML estimation of the parameter ϕ is of course no longer available. However, the Pearson estimator (3.9) may still be used—again, see McCullagh [McC83].

To summarize, the estimating equations approach is useful in situations where one can come up with reasonable assumptions concerning means and variances, but it is difficult to motivate the use of a particular distribution. In the absence of a likelihood, the idea is to start with a general class of estimators defined as solutions to equations like (3.42). The variance assumption then leads to the choice of a certain estimator within this class, by the criterion of asymptotic efficiency.

3.5.2 The Overdispersed Poisson Model

Including overdispersion into claim frequency models is a good example of the usefulness of the estimating equations approach. Starting from the variance assumption $V[Y_i] = \phi\mu_i/w_i$, in Sect. 3.4 we constructed a negative binomial model with this property. However, it was noted that the main reason for choosing this particular distribution is its analytical simplicity. General results ensure that the ML estimators are asymptotically efficient and normally distributed, *provided the distributional assumption holds true*. If this is not so, there is no guarantee that the asymptotic properties hold. A reason for using (3.39) in this case is thus one of *robustness*: the asymptotic results hold provided only that the variance assumption is true, regardless of the underlying distribution.

For different choices of the function $v(\mu)$, we get different models. The model corresponding to $v(\mu) = \mu$ is usually called the *overdispersed Poisson model* (ODP). This model, which does not assume a specific distribution, includes a dispersion parameter, which the ordinary Poisson model lacks. The ODP is a standard model for analyzing claim frequencies in non-life insurance.

3.5.3 Defining Deviances from Variance Functions*

An interesting property of the estimators derived as solutions to the estimating equations (3.39) is that they are so called *M-estimators*, see e.g. Andersen [An07]. M-estimators are obtained as the parameter values $\boldsymbol{\beta} = (\beta_1, \ldots, \beta_r)$ minimizing an expression of the type $\sum_i \rho(y_i, \boldsymbol{\beta})$, where $\rho(\cdot,\cdot)$ is a function which in some sense measures the fit to the data. For instance, least squares estimators are M-estimators with $\rho(y_i, \boldsymbol{\beta}) = w_i(y_i - \mu_i(\boldsymbol{\beta}))^2/2$. We shall now verify that the estimators defined through (3.39) are M-estimators, for general $g(\mu)$ and $v(\mu)$.

Let b be any twice differentiable function, such that b'' is positive, and let h denote the inverse of b'. Then, as was seen in Sect. 3.1, the function

$$d(y,\mu) = 2[yh(y) - b(h(y)) - yh(\mu) + b(h(\mu))], \tag{3.43}$$

has the properties that for fixed y, it is decreasing for $\mu < y$, zero for $\mu = y$, and increasing for $\mu > y$. Thus $d(y, \mu)$ is larger the further away from y the value μ is. Now, define $\rho(y_i, \boldsymbol{\beta}) = w_i d(y_i, \mu_i(\boldsymbol{\beta}))$ and let $\hat{\boldsymbol{\beta}}$ be the $\boldsymbol{\beta}$ minimizing $\sum_i \rho(y_i, \boldsymbol{\beta})$. Differentiating with respect to β_j for $j = 1, \ldots, r$ and setting the derivatives equal to zero, we get by (3.6),

$$\sum_i w_i \frac{h'(\mu_i)(y_i - \mu_i)}{g'(\mu_i)} x_{ij} = 0; \quad j = 1, \ldots, r.$$

Here we have omitted the constant 2. Comparing with (3.39), we see that if starting from $v(\mu)$ we can find a function b such that $h'(\mu) = 1/v(\mu)$ then we have shown that the solutions to (3.39) are M-estimators. With $v(\mu) = \mu$ we get $h(\mu) = \log \mu$, giving $b'(\theta) = \mathrm{e}^{\theta}$ and $b(\theta) = \mathrm{e}^{\theta}$. Similarly $v(\mu) = \mu^2$ leads to $b(\theta) = -\log(-\theta)$. The construction of b from v is in principle always possible, but of course the existence of an explicit expression requires that v is sufficiently simple.

We conclude that estimation using (3.39) may be viewed as a generalization of the method of least squares, with more general functions measuring the distance between the observations and the means. In the sequel, we will continue to call $d(y, \mu)$ in (3.43) a deviance, regardless of whether it is derived from a likelihood or a variance function.

Another way to view estimation without distributional assumptions is through the concept of *quasi-likelihood*, see e.g. Lee et al. [LNP06, Chap. 3].

3.6 Miscellanea

We have now described the basics of tariff analyses with GLMs, but there are of course a lot of additional practical issues. In this section we discuss some of them briefly.

3.6.1 Model Selection

In practice there is often a large number of possible models to choose among. In our analysis of moped insurance for instance, we could have gone further and tested the incorporation of new variables in the tariff, such as the owner's gender or age.

A problem with testing a number of possible variables for inclusion in any regression model—and tariff analysis is no exception—is that the more parameters we include in the model, the better the fit to the data. If we use the maximum number of parameters, $r = n$, we get a perfect fit; does this mean that we always should include as many rating factors as possible? The answer is of course no: we do not really look for a model that fits the data perfectly, but rather one that is good for predicting next year's outcome. This means that we are looking for variables with a statistically significant influence on the claim cost that is stable over time. The larger the model, the higher the risk for "nonsense-correlation" and a model that does not have real predictive power.

It should further be noted that a variable could have a statistically significant influence on the claim cost without being significant in the every-day use of the word, i.e. without being important. It is important to check whether the variable really makes a non-negligible difference in the price the customers would pay.

In general, one might say that statistical model selection should be based on the principle of parsimony, often referred to as Occam's razor, which in one of its Latin forms is stated as "pluralitas non est ponenda sine neccesitate." The approximate meaning of this is that "things should not be multiplied unnecessarily". In our case, the interpretation is that out of a number of plausible models, one should choose the simplest one.

Occam's razor:
Pluralitas non est ponenda sine neccesitate.
(Things should not be multiplied unnecessarily.)

3.6.2 Interaction

In a multiplicative model for a key ratio, any rating factor has the same relative impact, regardless of the value of the other rating factors. A case where multiplicativity is not a reasonable model is for gender and age in private car *third party liability* (TPL). Young men have a substantially higher claim frequency and severity than young women, while there is little gender difference among middle-aged drivers. This means that there is *interaction* between the rating factors age and gender. To include this effect in the model, we introduce a new factor, *gender/age*, which has levels like: men aged 18–29; women aged 18–29; men aged 30–39; women aged 30–39; etc. Table 3.6 exhibits a constructed example of what the price relativities could look like. As for other rating factors, the principle of *Ceteris paribus* applies, so that these interactive effects are the same in large cities as in rural areas, for example.

To be on the safe side, one might be tempted to introduce interactions for many of the rating factors, but Occam's razor tells us to avoid interactions unless they are absolutely necessary. A model with many interactions will be harder to interpret and lead to a large number of parameters, with the risk of estimates becoming very volatile.

Table 3.6 Tentative relativities for an *age/gender* interaction rating factor

	Age 18–29	Age 30–39	Age 40–49	...
Men	2.0	1.3	1.0	...
Women	1.2	1.1	1.0	...

3.6.3 Offsets

Sometimes, part of the expected value μ is known beforehand: it might be that relativities for some particular rating factor are taken from a special analysis, that some factors are set by the market department or even that mandatory values for all insurance companies are given by the authorities. This situation can be handled by including the corresponding factor as an *offset* in the GLM analysis. This is done on the linear scale: if we use a multiplicative model and the known relativities are given by the variable u, then we use $z_i = \log(u_i)$ as an offset as follows

$$\eta_i = z_i + \sum_{j=1}^{r} x_{ij}\beta_j \quad i = 1, 2, \ldots, n.$$

In the particular case of multiplicative Tweedie models, the ML equations in (2.30), supplemented by (2.31), becomes

$$\sum_i w_i \frac{y_i - u_i\mu_i}{(u_i\mu_i)^{p-1}} x_{ij} = 0 \iff \sum_i w_i u_i^{2-p} \frac{y_i/u_i - \mu_i}{\mu_i^{p-1}} x_{ij} = 0. \tag{3.44}$$

So in this important special case, offsetting $z_i = \log(u_i)$ is equivalent to using a GLM with observations y_i/u_i and weights $w_i u_i^{2-p}$.

3.6.4 Polynomial Regression

For continuous rating factors, like *policyholder age* or *car weight*, our approach so far has been to group the possible values of the variable into intervals, treating values in the same interval as identical. Another possibility is to use so called *polynomial regression*—to model the effect as a polynomial function. If x_i is the value of a continuous variable for observation i, this is achieved by including a term $\beta_k x_i + \beta_{k+1} x_i^2 + \cdots + \beta_{k+m-1} x_i^m$ into η_i in (2.19). In other words, we include the rating factors $x_i, x_i^2, \ldots, x_i^m$ into the model.

A main problem with polynomial regression is choosing the degree of the polynomial. A low degree may give a poor fit, while a high degree can lead to overfitting. The general technique of cross-validation, which we shall encounter in Sect. 5.5, provides one possible way of choosing the degree. In Chap. 5 we shall consider more sophisticated methods for modeling the effects of continuous variables.

3.6.5 Large Claims

For some insurance portfolios, a substantial part of the total claim cost may be due to a few large claims. Such dominating claims can make estimates very volatile and their effect has to be reduced somehow.

A common approach is to truncate the claims, i.e., to leave out the part of the claim cost that lies above some threshold $c > 0$. If a single claim cost is denoted by X_k, we define new variables $\tilde{X}_k \doteq \min(X_k, c)$, which replace X_k in the further analysis. Note that we do not truncate the average claim cost in the tariff cells, but the individual claims that are aggregated to that cost. This approach raises two problems: how to choose c and what to do with the omitted (truncated) claim cost.

The threshold c is usually determined by trial and error. The goal is, on the one hand, to choose c as large as possible, in order for the analysis to be relevant; on the other hand, c has to be chosen small enough to give reliable estimates. The compromise between these two will have to be based on subjective judgement.

The truncated cost—the cost above the threshold—also has to be provided for by the premiums. A common solution is to distribute this cost among the policyholders by the relativities found in the GLM analysis of the truncated $\tilde{X}_k$. In practice, this means that the relativities from the analysis of the $\tilde{X}$'s are left unchanged and one only adjusts the base premium so that the overall premium level is adequate. This approach implicitly assumes that the relativities for large claims are the same as those for the rest, the so called *attritional losses*.

However, it may well be that some groups of policyholders contribute more to the large claims than indicated by the attritional losses. It goes without saying that we cannot perform a complete GLM analysis of claim frequency and severity for the large losses—if so, truncation would not be necessary. Notwithstanding this fact, the *proportion* of large claims might be stable enough to allow reliable estimation. This can be performed by a binomial GLM, see Exercise 2.6, with the total number of claims as exposure and the number of large claims as response; the logit-link in (2.21) is appropriate for this analysis. Now the large claim frequency can be calculated as the original claim frequency times the proportion of large claims. Unfortunately, this approach has some problems left, one being that we do not have a claim severity model for large claims but will have to use some average value, another is that when the results for attritional losses and large claims are added, the resulting tariff will no longer be multiplicative.

3.6.6 Deductibles*

A deductible (in the UK called an excess) is the part of the damages to be covered by the policyholder. Often this is limited to a fixed amount; if the deductible is a maximum EUR 200, then the policyholder is not compensated for costs up to that value, while in the case of a larger claim, the loss of the policyholder is restricted to EUR 200 and the rest is paid by the insurance company.

If the insurance policy defines a fixed deductable for all policyholders, the analysis can simply be carried out net of the deductibles. If the level of deductible has increased over the years, we may adjust the data *as if* today's deductible had been in force over the entire data collecting period. The more uncommon situation with a decreased level is harder, since claims with a cost between the old and new level may not be registered. As with other deficiencies in the data there is no general remedy for this problem; one simply has to use common sense.

In the case where the customer can choose between two or more deductible levels, other problems arise. One might think of using the level of deductible as a rating factor of its own. This is, however, not a good idea, as will now be demonstrated.

Suppose there are two possible choices, EUR 200 and EUR 400. Suppose we have two customers, one with a mean claim severity of EUR 500 before withdrawing the deductible and another with EUR 1 000. The net mean severity is then $500 - 200 = 300$ and $500 - 400 = 100$ for the first customer with the respective levels of deductible. For the second customer, the corresponding amounts are 800 and 600, respectively. The ratios between these amounts for the two customers are then $300/100 = 3.0$ and $800/600 = 1.3$, respectively. If deductible had been a multiplicative rating factor, these ratios should have been the same. In practice, it is inappropriate to treat deductible as a multiplicative rating factor if the claim severity varies significantly between the tariff cells.

A possible way to handle this problem is to make separate models for each *layer* of the claim. If we have just the two choices of deductible EUR 200 and EUR 400, we have two layers: one between 200 and 400 and one over 400. For the second layer, we create a data set with all policies, but only the claims above 400 and for these only that part of the damages that exceeds 400. For the first layer, we create a data set consisting only of policies having the lower deductible, and only that part of the claims that fall between 200 and 400. We then analyze these two data sets separately, using a standard GLM model for each one. The analysis for the second layer yields a pure premium that has to be paid by all policyholders; the one for the first layer yields the additional premium to be paid by those who choose the lower deductible.

While this model is reasonable, and makes effective use of all the data, we might run into trouble since this kind of tariff is not multiplicative when the two layers are added. In effect, the layers are to be treated as different parts of the cover and if we have several possible choices of deductible, as is common in the commercial lines, this method may become intractable.

3.6.7 Determining the Premium Level

As stated in the first chapter, the goal of a tariff analysis is to assess the *relative* premium levels to be charged to the policyholders in the tariff cells. The main output is the relativities and we seldom make use of the base premium that comes out of the GLM analysis.

Setting the tariff's base premium involves considerations outside the scope of the GLM analysis. Since this contains little of statistical interest we shall not go into details on this matter. Here are just a few words for the sake of completeness.

The calculation is often tantamount to making a budget. The starting point may be a forecast of the risk premium, taking into account expected changes in the portfolio, inflation and reinsurance. This gives an expected claim cost for the entire portfolio, to which is added claims handling costs and operating expenses. Investment return on the technical provisions (for outstanding claims and premiums) as well as a profit loading should also be taken into account. Finally, the choice of premium level is affected by market considerations.

3.7 Case Study: Model Selection in MC Insurance

This case study is a direct continuation of Case study Sect. 2.4. The task here is to determine a new tariff for the MC insurance.

Problem 1: Extend the analysis of claim frequency and claim severity in Case study Sect. 2.4 to include confidence intervals and p-values for the rating factors in a "Type 3 analysis". Use the Pearson estimator of ϕ.
Problem 2: Determine a preliminary new tariff based on the rating factors in the old tariff. This might include the exclusion of some rating factors and/or the merging of some classes. Make the changes to claim frequency and claim severity separately, i.e., you may exclude a rating factor for one of them while retaining it for the other. Compare your result to the old tariff.
Problem 3: Next have a look at previously unused variables, the gender and age of the owner, to see if one or both have a significant effect.
Problem 4: In case of significance, examine whether there is any interaction between the two new rating factors. (Other interactions might exist but need not be investigated in this case study.)
Problem 5: Summarize all the analyses above to give the actuary's proposal for a new tariff. Compare your result to the old tariff.

Exercises

3.1 (Section 3.1) Show that if we have a GLM with normal distribution and all $w_i = 1$ then the deviance D is given by (3.3).

3.2 (Section 3.1) Find an expression for the deviance D in the case of the gamma model in (2.5).

3.3 (Section 3.1) For Tweedie models with $1 < p < 2$ and $p > 2$, (2.10) reads

$$b(\theta) = -\frac{1}{p-2}(-(p-1)\theta)^{(p-2)/(p-1)}.$$

Find an expression of the deviance for these models.

3.4 (Section 3.1) In the discussion of Fig. 3.1, it was suggested that it may make little difference which of our standard deviances you choose. For the data in the case study on motorcycle insurance in Sect. 2.4, try estimating the relativities assuming a normal distribution instead of a Poisson or gamma distribution and compare the result.

3.5 (Section 3.1)

(a) For the Poisson case, show that the unscaled deviance D is approximately equal to the unscaled Pearson X^2, by a Taylor expansion of (3.7) as a function of y_i/μ_i around 1.
(b) Repeat (a) for the gamma case.

3.6 (Section 3.1) Derive the LRT statistic for the gamma claim severity model. *Hint*: use Exercise 3.2.

3.7 (Section 3.2) Verify the expression in (3.12).

3.8 (Section 3.2) Verify the expression in (3.15).

3.9 (Section 3.2) Construct the confidence intervals in Table 3.4 from Tables 3.2 and 3.3, without using any other output from the separate analysis of F and S.

3.10 (Section 3.2) Show that in the case of canonical link we have $g'(\mu_i) = 1/v(\mu_i)$ and that this implies $\mathbf{A} = \mathbf{D}$, and hence $\mathbf{I} = -\mathbf{H}$.

3.11 (Section 3.2) With $s_j(\mathbf{y}, \boldsymbol{\beta}, \phi)$ as in (3.22), let $\mathbf{V}$ be the matrix with elements $v_{jk} = \mathrm{Cov}(s_j(\mathbf{Y}, \boldsymbol{\beta}^0, \phi), s_k(\mathbf{Y}, \boldsymbol{\beta}^0, \phi))$. Show that $\mathbf{V} = \mathbf{I}$, with $\mathbf{I}$ as in (3.14).

3.12 (Section 3.3) For the relative Gamma distribution for claim severity in (2.5), determine the residuals r_{Pi} and r_{Di}.

3.13 (Section 3.4)

(a) Prove (3.37) using the law of total probability (Lemma A.4).
(b) Show that the moment generating function of the distribution (3.37) is

$$M(t) = (1 - \nu(e^t - 1))^{-w_i \mu_i / \nu}. \tag{3.45}$$

What can be said about the sum of two independent variables with the distribution (3.37)?

3.14 (Section 3.4) In this exercise, assume (3.33).

(a) Using that $\Gamma(t+1) = t\Gamma(t)$, show the loglikelihood of the observed claim frequencies $\{y_i\}$ may be written

$$\ell(\mathbf{y}; \boldsymbol{\mu}, \nu) = \sum_i \log f_{Y_i}(y_i; \mu_i, \nu)$$

$$= \sum_i \left\{ \sum_{k=1}^{w_i y_i} \log(w_i \mu_i + (k-1)\nu) \right.$$

$$\left. - \left(\frac{w_i \mu_i}{\nu} + w_i y_i \right) \log(1+\nu) - \log((w_i y_i)!) \right\}. \quad (3.46)$$

(b) Prove that with the usual multiplicative model, i.e. $\mu_i = \exp\{\eta_i\}$, with $\eta_i = \beta_1 x_{i1} + \cdots + \beta_r x_{ir}$, the ML equation for β_j becomes

$$\sum_i w_i \sum_{k=1}^{w_i y_i} \left(\frac{\mu_i \nu}{w_i \mu_i + (k-1)\nu} - \mu_i \log(1+\nu) \right) x_{ij} = 0$$

and that the ML equation for ν is

$$\sum_i \left\{ \sum_{k=1} \frac{k-1}{w_i \mu_i + (k-1)\nu} + \frac{w_i \mu_i}{\nu^2} (\log(1+\nu) - \nu) + \frac{w_i \mu_i - w_i y_i}{1+\nu} \right\} = 0.$$

3.15 (Section 3.4) Let N be a random variable taking values on the non-negative integers $0, 1, \ldots$ and put $p_n = P(N = n)$. The probability generating function of N is

$$P(s) = \sum_{n=0}^{\infty} p_n s^n.$$

If $P^{(k)}(s)$ denotes $d^k P(s)/ds^k$, then

$$P^{(k)}(0) = k! p_k. \quad (3.47)$$

(a) Assuming that N has the cumulant generating function (2.8), derive the probability generating function of N.
(b) Use (3.47) to conclude that if ϕ is not an integer, then there is no distribution on the non-negative integers with a probability mass function of the form (2.1).

3.16 (Section 3.4) In this exercise, assume (3.35).

(a) Prove (3.38) using the law of total probability (Lemma A.4).
(b) Show that with the usual multiplicative model, the ML equation for β_j is

$$\sum_i \left\{ w_i y_i - w_i \mu_i \frac{w_i y_i + 1/\nu}{w_i \mu_i + 1/\nu} \right\} x_{ij} = 0.$$

(c) Show that if ν is considered known, then (3.38) may be viewed as an EDM with $h(\mu) = \log(\nu w_i \mu/(1 + \nu w_i \mu))$, where h is the inverse of b'.

3.17 (Section 3.5) A linear estimator of the parameters $\beta_1, \ldots, \beta_r$ has the form $\hat{\beta}_j = \sum_i c_{ji} y_i$, $j = 1, \ldots, r$. Show that any unbiased linear estimator may be obtained as the solution to equations of the form (3.41).

3.18 (Section 3.5) Starting from a two-parametric distribution, it is in principle always possible to produce a model satisfying the variance assumption $V[Y_i] = \phi v(\mu_i)/w_i$. Here is an example for claim severity. Assume that the size X_i of claim i satisfies

$$P(X_i > x) = \left(\frac{(\gamma - 1)\mu_i}{(\gamma - 1)\mu_i + x} \right)^{\gamma}, \quad x > 0.$$

This is the generalized Pareto distribution. Show that $E[X_i] = \mu_i$ and $V[X_i] = \phi \mu_i^2$, with $\phi = \gamma/(\gamma - 2)$. (The model is defined only for $\gamma > 2$.)

Chapter 4
Multi-Level Factors and Credibility Theory

So far we have studied how GLMs may be used to estimate relativities for rating factors with a moderate number of levels: the rating factors were either categorical with few levels (e.g. *gender*) or formed by a grouping of a continuous variable (e.g. *age group*, *mileage class*). In case of insufficient data, one can merge levels to get more reliable estimates, nota bene at the price of a less diversified tariff: the age group 20–29 is not as homogenous as the group 20–24 or 25–29, but merging may be a way out of the problems caused by lack of data in one or both of the latter groups. However, for categorical rating factors with a large number of levels without an inherent ordering, there is no simple way to form groups to solve the problem with insufficient data. We introduce the term *multi-level factor* (MLF) for such rating factors; here are some examples.

Example 4.1 (Car model classification) In private motor insurance it is well known that the car model is an important rating factor; exclusive cars, for instance, are more attractive to thieves and more expensive to replace than common cars. In Sweden there are some 2 500 car model codes, some of which represent popular cars with sufficient data available, whereas most have moderate or sparse data. There is no obvious way to group the model codes a priori and there is not enough data to estimate 2 500 GLM parameters. Hence, car model is a typical MLF.

Example 4.2 (Experience rating) Using the *customer* as a rating factor is another important example of an MLF. In the commercial lines it is common to base the rating to some extent on the individual claims experience, even though we usually do not have enough data for a completely separate rating of each company. This is the classical situation for which North-American actuaries in the early 1900's introduced *credibility estimators*, which are weighted averages of (i) the pure premium based on the individual claims experience; and (ii) the pure premium for the entire portfolio of insurance policies. Another form of experience rating is represented by the bonus/malus systems that are common in private motor TPL insurance; here again the customer can be regarded as an MLF.

E. Ohlsson, B. Johansson, *Non-Life Insurance Pricing with Generalized Linear Models*, EAA Lecture Notes,

DOI 10.1007/978-3-642-10791-7_4,

Example 4.3 (Geographic zones) A problem of the same type as car model classification is how the premium should be affected by the area of residence of the customer. In order to get risk homogenous geographic areas one often has to use a very fine subdivision of the country, based on, for instance, postcodes. We might have several thousand such areas and obviously can not use postcode as an ordinary GLM rating factor. Of course, we can easily form larger groups with enough data by merging neighboring areas; however, these can have quite different risk profiles: a wealthy residential area is probably better matched to similar areas from other cities than to a neighboring low-income area. In any case, a prior grouping can be very hard to achieve and so postcode, or any fine geographic segmentation, can be seen as an MLF. A possibility is of course to use information on, say, household income, population density or age of the residents; such information may be supplied by the national statistical agency. In our experience, though, the available variables can not explain all the risk variation between geographic areas, and we are still left with an MLF.

These three examples have in common that the variable of interest is categorical, with so many categories that we do not have enough data to get reliable estimates for each category. Furthermore, there is no obvious ordering of the categories: we have a *nominal* variable. So a multi-level factor (MLF) is in short a nominal variable with too many categories for ordinary GLM estimation. Admittedly, this definition is a bit vague: it is up to the actuary to judge which variables are to be treated as MLFs in a particular tariff analysis. It should be noted that we might have sufficient data for *some* levels of an MLF, while for others data is scarce.

The branch of insurance mathematics where problems of this kind are studied is called *credibility theory*. Here we have an MLF (in our terminology), for which we are interested in some key ratio Y, such as the pure premium. We have observations Y_{jt}, with exposure weights w_{jt}, where j is the MLF level and t denotes repeated observations within the level. The w-weighted average $\overline{Y}_{j\cdot} = (\sum_t w_{jt} Y_{jt})/(\sum_t w_{jt})$ is an estimate of the individual key ratio. The idea of credibility theory is to use a compromise between the more relevant but unstable $\overline{Y}_{j\cdot}$, and the average for the entire portfolio, μ, in form of a *credibility estimator*

$$z_j \overline{Y}_{j\cdot} + (1 - z_j)\mu, \tag{4.1}$$

where μ is estimated somehow, often by the weighted total average

$$\overline{Y}_{\cdot\cdot} = \frac{\sum_j w_{j\cdot} \overline{Y}_{j\cdot}}{\sum_j w_{j\cdot}}, \tag{4.2}$$

which is more stable than $\overline{Y}_{j\cdot}$.

The number z_j, $0 \le z_j \le 1$, is called a *credibility factor* (or credibility weight), and measures the amount of credibility we give to the individual premium.

Already in the early 1900's, American actuaries where discussing how the credibility factors in (4.1) should be chosen. Assume for concreteness that the levels of the MLF correspond to different companies. It is clear that z_j should be larger the more data we have on company j and that it should be larger the less the variation in the company data. Finally, we should give more credibility to the company's data if there is much variation between the companies—in which case $\overline{Y}_{\cdot\cdot}$ has little to say about the risk level of a particular company.

Modern credibility theory dates back to the 60's, when European insurance mathematicians showed how credibility estimators like the one in (4.1) can be obtained from statistical models: some important early papers are [Bü67, BS70, Je74]. We introduce the ideas by giving an example.

Example 4.4 (Bonus/malus in Swiss motor TPL) We have already encountered experience rating in Example 4.2; the present example concerns the Swiss bonus system for private motor TPL insurance; it is based on the pioneering work by Bichsel from 1964, recapitulated in Goovaerts and Hoogstad [GH87, p. 35]. The idea is to base the claim frequency estimate partly on the claims experience of the individual customer, indexed by j. If w_{jt} is the duration of the policy during year t, and N_{jt} is the number of claims caused by the customer during that period, then $w_{j\cdot} = \sum_t w_{jt}$ and $N_{j\cdot} = \sum_t N_{jt}$ are the total exposure to risk and the total number of claims that we observe for that customer, respectively. The empirical claim frequency is then $\overline{Y}_{j\cdot} = N_{j\cdot}/w_{j\cdot}$.

It is a reasonable assumption that N_{jt} follows a Poisson distribution with expectation $w_{jt}\lambda_j$ for some λ_j; then $N_{j\cdot} \sim Po(w_{j\cdot}\lambda_j)$. Note that $E(\overline{Y}_{j\cdot}) = \lambda_j$, and so λ_j is the expected annual claims frequency. Bichsel's idea was to consider λ_j as the outcome of a random variable Λ_j. The Λ_j's are assumed to be independent and identically distributed; their common distribution describes the variation in *accident proneness* amongst the customers: each customer has an individual expected claims frequency drawn from this distribution. The insurance company does not know λ_j, and the claims history of the customer is not large enough to get a reliable estimate. The idea with a bonus system is that the claims history should, nevertheless, influence the premium, since few claims in the past can be taken as an indication that the customer is one with a low, albeit unknown, λ_j.

Now, if we can not observe the outcome of Λ_j, its expectation $\mu = E(\Lambda_j)$ is a natural quantity to use for rating, and for a new customer we can not do much more than this. For an old customer, with claims history summarized by $N_{j\cdot}$, it seems better to use $E(\Lambda_j|N_{j\cdot})$. If we assume that Λ_j is Gamma distributed, $G(\alpha, \beta)$ in the parametrization of (2.4), then $\mu = \alpha/\beta$, and it may be shown that

$$E(\Lambda_j|N_{j\cdot}) = z_j \frac{N_{j\cdot}}{w_{j\cdot}} + (1 - z_j)\mu, \quad \text{with } z_j = \frac{w_{j\cdot}}{w_{j\cdot} + \beta}. \tag{4.3}$$

Since $\overline{Y}_{j\cdot} = N_{j\cdot}/w_{j\cdot}$, this is just an instance of (4.1). The point is that we now have an expression for z_j; if we can estimate α and β—which should be possible since they are the same for all costumers—we have found a way to compute a credibility estimator. The larger $w_{j\cdot}$, i.e., the more we know about the customer, the more credibility z_j we give to the claims history $\overline{Y}_{j\cdot}$.

We shall refrain from a proof of (4.3) here, since it is merely a special case of the results below.

4.1 The Bühlmann-Straub Model

We begin by deriving the classical Bühlmann-Straub estimator, which concerns the case with one single MLF and no other rating factors. We use a notation suitable for the generalization to the case with an MLF plus ordinary GLM rating factors which is the topic of the following sections.

In Example 4.4 the parameter corresponding to the MLF was modeled as a random variable. The reader may be familiar with this idea from the analysis of variance (ANOVA), where such *random effects* appear, i.e., in variance component models. The concept of random parameters is the basis of the derivation of modern credibility estimators.

As opposed to Example 4.4, the Bühlmann-Straub model only makes assumptions on the first two moments—not the entire distribution of the random variables. This is an advantage in applications, where we might have an opinion on the variance even if we do not know the entire distribution. Unless otherwise stated, we assume below that the first two moments of all random variables exist.

We keep the notation from the previous section, where Y_{jt} is some key ratio for MLF level j and observation t, and μ is the overall mean. Here we will talk about *group* j, when considering all policies that have level j on the MLF. We let u_j be the relativity for that group; $j = 1, 2, \ldots J$. With sufficient data, the u_j's could have been estimated by standard GLM techniques, but in the case of an MLF, one or more of the groups do not have enough observations for reliable estimation, but we still want to make the best use of the information we have. Credibility estimators for this case can be derived under the assumption that relativities are (observations of) random effects U_j. The basic multiplicative model is then

$$E(Y_{jt}|U_j) = \mu U_j, \tag{4.4}$$

where $E(U_j) = 1$, in order to avoid redundancy in the parameters.

Even though relativities are standard in GLM rating, it will be more convenient for our derivations to work with the variable $V_j \doteq \mu U_j$, rather than U_j. By assumption, $E(V_j) = \mu$ and

$$E(Y_{jt}|V_j) = V_j. \tag{4.5}$$

In rating with GLMs, we mostly assume that the data follows a *Tweedie model*, with the variance proportional to the pth power of the expectation, see Sect. 2.1.4. In our case we must condition on the random effect; the Tweedie variance is then, using Lemma 2.1 and (2.9),

$$\text{Var}\left(Y_{jt}|V_j\right) = \frac{\phi V_j^p}{w_{jt}}, \tag{4.6}$$

where ϕ, is the GLM dispersion parameter.

Assuming that the V_j are identically distributed, we may introduce $\sigma^2 \doteq \phi E[V_j^p]$, which is independent of j. From (4.6) we then have

$$E\left[\operatorname{Var}\left(Y_{jt}|V_j\right)\right] = \frac{\sigma^2}{w_{jt}}. \tag{4.7}$$

This is really the only property of the variance of Y_{jt} that we need here. We now collect our assumptions for the basic credibility model, recalling that all random variables are assumed to have finite second moment.

Assumption 4.1

(a) The groups, i.e., the random vectors $(V_j, Y_{j1}, Y_{j2}, Y_{j3}, \ldots)$ are independent for $j = 1, \ldots, J$.
(b) The V_j; $j = 1, 2, \ldots, J$; are identically distributed with $E[V_j] = \mu > 0$ and $\operatorname{Var}(V_j) = \tau^2$ for some $\tau^2 > 0$.
(c) For any j, conditional on V_j, the Y_{jt}'s are mutually independent, with mean given by (4.5) and variance satisfying (4.7).

Note that we do not assume a Tweedie model; these models were only used to motivate (4.7). Note also that

$$\operatorname{Var}\left(Y_{jt}\right) = \operatorname{Var}[E(Y_{jt}|V_j)] + E[\operatorname{Var}(Y_{jt}|V_j)] = \tau^2 + \frac{\sigma^2}{w_{jt}}, \tag{4.8}$$

a structure familiar from variance component models.

With enough data, we would treat the rating factor as a fixed effect and an obvious estimator of V_j would be the weighted mean

$$\overline{Y}_{j\cdot} = \frac{\sum_t w_{jt} Y_{jt}}{\sum_t w_{jt}}. \tag{4.9}$$

If, on the other hand, there where no data at all for group j, a reasonable key ratio would be (an estimate of) μ—a credibility estimator should be a compromise between these two extremes. Formally, we define the credibility estimator of a random effect V as the linear function $\widehat{V}$ of the observations $\mathbf{Y}$ that minimizes the mean square error of prediction (MSEP),

$$E\left[(h(\mathbf{Y}) - V)^2\right] \tag{4.10}$$

among all linear functions $h(\mathbf{Y})$.

Remark 4.1 Some might prefer to call $\widehat{V}$ a predictor rather than an estimator, since V is a random variable and not a parameter, and it is sometimes called the BLP (best linear predictor). The actuarial tradition is, however, to call it an estimator, and we adhere to this terminology, noting that V is not something that will occur in the future—rather, it is a random variable whose outcome no one will ever be able to observe. For further discussion, in the general statistical area, see Robinson [Ro91, Sect. 7.1].

The solution to the minimization problem in (4.10) is the famous estimator of Bühlmann and Straub [BS70]. Note, though, that our notation is not the classical one, cf. Sect. 4.1.2.

Theorem 4.1 (Bühlmann-Straub) *Under Assumption* 4.1, *the credibility estimator of V_j is given by*

$$\widehat{V}_j = z_j \overline{Y}_{j\cdot} + (1 - z_j)\mu, \tag{4.11}$$

where

$$z_j = \frac{w_{j\cdot}}{w_{j\cdot} + \sigma^2/\tau^2}. \tag{4.12}$$

Here μ might be given a priori, e.g. by a tariff value; otherwise it could be estimated by the weighted mean $\overline{Y}_{\cdot\cdot}$ in (4.2). Estimators of τ^2 and σ^2 are given in the next section. We observe that a credibility estimator of the random effect U_j can be obtained as $\widehat{U}_j = z_j \overline{Y}_{j\cdot}/\mu + (1 - z_j)$.

In order to prove the theorem, we start by restating a result by Sundt [Su80].

Lemma 4.1 *A linear estimator $\widehat{V}$ of V, based on* **Y**, *is the credibility estimator, if and only if,*

$$E(\widehat{V}) = E(V); \tag{4.13}$$

$$\text{Cov}(\widehat{V}, Y_t) = \text{Cov}(V, Y_t); \quad t = 1, 2, \ldots. \tag{4.14}$$

Sundt also refers to a result by de Vylder, by which there always exists a unique credibility estimator; hence it is correct to speak of "*the* credibility estimator".

Proof We shall prove the "if" statement only, which is the part that we need in the proof of the theorem. For a proof of the "only if" part, see Sundt [Su80]. So we shall prove that if $\hat{h}$ fulfills (4.13) and (4.14), then it minimizes the MSEP in (4.10). Now let $h = h(\mathbf{Y}) = c_0 + \sum_t c_t Y_t$ be any linear estimator. Its MSEP can be developed as follows,

$$\begin{aligned} E[(h - V)^2] &= E[(h - \widehat{V} + \widehat{V} - V)^2] \\ &= \text{Var}(h - \widehat{V} + \widehat{V} - V) + [E(h - \widehat{V} + \widehat{V} - V)]^2 \\ &= \text{Var}(h - \widehat{V}) + \text{Var}(\widehat{V} - V) + 2\,\text{Cov}[(h - \widehat{V}), (\widehat{V} - V)] \\ &\quad + [E(h) - E(V)]^2. \end{aligned}$$

But $h - \widehat{V}$ is a linear combination of the Y's, say $d_0 + \sum_t d_t Y_t$ and hence the covariance above can be expanded as

$$\sum_t d_t \,\text{Cov}[Y_t, (\widehat{V} - V)] = \sum_t d_t [\text{Cov}(Y_t, \widehat{V}) - \text{Cov}(Y_t, V)],$$

which is equal to zero by (4.14), so that the optimal choice of h is that which minimizes

$$E[(h-V)^2] = \text{Var}(h-\widehat{V}) + \text{Var}(\widehat{V}-V) + [E(h)-E(V)]^2 \geq \text{Var}(\widehat{V}-V).$$

The right-most member is unaffected by the choice of h and is hence a lower bound for the MSEP. If we let $h=\widehat{V}$ the first term in the middle member becomes zero, and so does the last one, since $E(\widehat{V})=E(V)$ by (4.13); thus the lower bound is achieved for this choice of h. This proves that if $\widehat{V}$ fulfills the assumptions of the lemma, it is the credibility estimator. □

Proof of Theorem 4.1 We shall prove that the equations of Lemma 4.1 are fulfilled for $\widehat{V}_j$ in (4.11). First, since $E(V_j)=\mu$, and $E(Y_{jt})=E[E(Y_{jt}|V_j)]=E(V_j)=\mu$, we have $E(\widehat{V}_j)=E(V_j)$ so that (4.13) is valid in this case.

In our case, (4.14) can be written as $\text{Cov}(\widehat{V}_j, Y_{j't}) = \text{Cov}(V_j, Y_{j't})$, for all j' and t. Because of the independence of groups in Assumption 4.1(a), we have $\text{Cov}(V_j, Y_{j't})=0$ for $j' \neq j$, and since $\widehat{V}_j$ only includes Y-values from group j, (4.14) is trivially fulfilled as soon as $j' \neq j$; thus we only have to consider the case $j'=j$ in the following.

We use Lemma A.3(b) to show that

$$\text{Cov}(V_j, Y_{jt}) = \text{Cov}[V_j, E(Y_{jt}|V_j)] = \text{Var}(V_j) = \tau^2.$$

For $s \neq t$ we have by Assumption 4.1(c), this time using part (a) of Lemma A.3,

$$\begin{aligned}\text{Cov}(Y_{js}, Y_{jt}) &= E[\text{Cov}(Y_{js}, Y_{jt}|V_j)] + \text{Cov}[E(Y_{js}|V_j), E(Y_{jt}|V_j)]\\ &= 0 + \text{Var}[V_j] = \tau^2.\end{aligned}$$

For $s=t$, on the other hand, we have by (4.8),

$$\text{Cov}(Y_{jt}, Y_{jt}) = \text{Var}(Y_{jt}) = \tau^2 + \frac{\sigma^2}{w_{jt}}.$$

Hence,

$$\begin{aligned}\text{Cov}(\widehat{V}_j, Y_{jt}) &= z_j \sum_s \frac{w_{js}}{w_{j\cdot}} \text{Cov}(Y_{js}, Y_{jt}) = z_j \left(\tau^2 + \frac{w_{jt}}{w_{j\cdot}} \frac{\sigma^2}{w_{jt}}\right)\\ &= \tau^2 z_j \left(\frac{w_{j\cdot} + \sigma^2/\tau^2}{w_{j\cdot}}\right) = \tau^2,\end{aligned}$$

by (4.12). This completes the proof of Theorem 4.1. □

4.1.1 Estimation of Variance Parameters

The above credibility estimation procedure reduces the number of parameters to be estimated from $J+1$, viz. $u_1, \ldots, u_J$ and ϕ, in the GLM case, to two (or three, depending on whether μ is given a priori or not). These two are σ^2 and τ^2, for which we next present estimators; the point is that we use the entire portfolio for these estimators, so the lack of data for some groups j is not a problem.

The parameter σ^2 in (4.7) can be regarded as a measure of the (weighted) within–group–variance. By analogy to ANOVA, an idea is to use the following sum of squares estimator of σ^2; see also Lemma A.6 in Appendix A. With n_j denoting the number of repeated observations at level j, let

$$\hat{\sigma}_j^2 = \frac{1}{n_j - 1} \sum_t w_{jt}(Y_{jt} - \overline{Y}_{j\cdot})^2,$$

which we shall see is an unbiased estimator of σ^2. We can increase accuracy by combining these estimators, weighted by the "degrees of freedom", to

$$\hat{\sigma}^2 = \frac{\sum_j (n_j - 1)\hat{\sigma}_j^2}{\sum_j (n_j - 1)}. \tag{4.15}$$

Similarly, $\tau^2 = \mathrm{Var}(V_j)$ is a measure of the between–group–variance and an estimator might be based on the sum of squares $\sum_j w_{j\cdot}(\overline{Y}_{j\cdot} - \overline{Y}_{\cdot\cdot})^2$. It turns out that we get an unbiased estimator by adjusting this sum of squares as follows,

$$\hat{\tau}^2 = \frac{\sum_j w_{j\cdot}(\overline{Y}_{j\cdot} - \overline{Y}_{\cdot\cdot})^2 - (J-1)\hat{\sigma}^2}{w_{\cdot\cdot} - \sum_j w_{j\cdot}^2 / w_{\cdot\cdot}}. \tag{4.16}$$

Theorem 4.2 *Under Assumption* 4.1, *the estimators* $\hat{\sigma}^2$ *and* $\hat{\tau}^2$ *are unbiased, i.e.,*

(a) $E[\hat{\sigma}^2] = \sigma^2$;
(b) $E[\hat{\tau}^2] = \tau^2$.

Before proving this theorem, we shall give a numerical example of credibility estimation, now that we have all the required components.

Example 4.5 (Group health insurance) For a collective insurance that gives compensation for income loss, consider a portfolio of five companies of moderate size operating in the same line of business. We want to estimate the pure premium per employee for the next year, based on the last ten years' (index adjusted) observations. Artificial data are given in Table 4.1 with observed pure premium and exposure measured in 10th of employees (so the first company had around 440 employees during the first year). The grand average of the pure premium is 481.3 and the question is how much the individual companies premium should deviate from

Table 4.1 Group health insurance: observed pure premium and 10th of employees (in brackets)

Year	Company 1	2	3	4	5
1	540 (44)	99 (20)	0 (8)	275 (22)	543 (26)
2	514 (50)	103 (20)	400 (6)	278 (22)	984 (24)
3	576 (56)	163 (24)	1042 (10)	430 (18)	727 (22)
4	483 (58)		313 (6)	196 (20)	562 (18)
5	481 (58)		0 (8)	667 (12)	722 (20)
6	493 (56)		833 (4)	185 (10)	610 (16)
7	438 (54)		0 (6)	517 (12)	794 (12)
8	588 (52)			204 (10)	299 (14)
9	541 (52)			323 (6)	580 (14)
10	441 (46)			968 (6)	
$\overline{Y}_{j\cdot}(w_{j\cdot})$	509.3 (526)	124.3 (64)	375.6 (48)	359.9 (138)	661.9 (166)

Table 4.2 Group health insurance: credibility factors and estimators; here $\hat{\mu} = \overline{Y}_{\cdot\cdot} = 481.3$

	Company 1	2	3	4	5
$w_{j\cdot}$	526	64	48	138	166
$\overline{Y}_{j\cdot}$	509.3	124.3	375.6	359.9	661.9
z_j	0.95	0.72	0.66	0.85	0.87
$\widehat{V}_j$	508.0	224.2	411.7	378.5	638.4
$\widehat{U}_j$	1.055	0.466	0.855	0.786	1.326

this average because of their own claims history. Company 2, for instance, will argue that they constitute a much lower risk than the average and should hence pay a smaller premium, but how far can we go in that direction based on three observations only?

We first use (4.15) and (4.16) to estimate $\hat{\sigma}^2 = (771.77)^2$ and $\hat{\tau}^2 = (154.72)^2$ and then compute the Bühlmann-Straub estimator according to (4.11). The results in Table 4.2 show rather high credibility factors, due to quite large variations between companies, as compared to those between the years, within company. Note that it is not all that easy to see this directly from the data.

Proof of Theorem 4.2 (a) Since $E(\overline{Y}_{j\cdot}|V_j) = E(Y_{jt}|V_j) = V_j$, using Lemma A.2,

$$E(Y_{jt} - \overline{Y}_{j\cdot})^2 = \text{Var}(Y_{jt} - \overline{Y}_{j\cdot}) = E[\text{Var}((Y_{jt} - \overline{Y}_{j\cdot})|V_j)].$$

By the conditional independence of the Y_{jt}'s given V_j,

$$\begin{aligned}\operatorname{Var}(Y_{jt} - \overline{Y}_{j\cdot} | V_j) &= \operatorname{Var}\left(\left(1 - \frac{w_{jt}}{w_{j\cdot}}\right) Y_{jt} - \sum_{t' \neq t} \frac{w_{jt'}}{w_{j\cdot}} Y_{jt'} \middle| V_j\right) \\ &= \left(1 - \frac{w_{jt}}{w_{j\cdot}}\right)^2 \operatorname{Var}(Y_{jt}|V_j) + \sum_{t' \neq t} \frac{w_{jt'}^2}{w_{j\cdot}^2} \operatorname{Var}(Y_{jt'}|V_j).\end{aligned}$$

Upon taking expectations and using (4.7), we conclude that

$$\begin{aligned}E(Y_{jt} - \overline{Y}_{j\cdot})^2 &= \left(1 - \frac{w_{jt}}{w_{j\cdot}}\right)^2 E[\operatorname{Var}(Y_{jt}|V_j)] + \sum_{t' \neq t} \frac{w_{jt'}^2}{w_{j\cdot}^2} E[\operatorname{Var}(Y_{jt'}|V_j)] \\ &= \frac{\sigma^2}{w_{jt}}\left(1 - 2\frac{w_{jt}}{w_{j\cdot}}\right) + \sum_{t'} \frac{w_{jt'}^2}{w_{j\cdot}^2} \frac{\sigma^2}{w_{jt'}} \\ &= \frac{\sigma^2}{w_{jt}}\left(1 - \frac{w_{jt}}{w_{j\cdot}}\right).\end{aligned}$$

Hence,

$$E\left[\sum_t w_{jt}(Y_{jt} - \overline{Y}_{j\cdot})^2\right] = \sigma^2 \sum_{t=1}^{n_j} \left(1 - \frac{w_{jt}}{w_{j\cdot}}\right) = \sigma^2 (n_j - 1).$$

This proves the unbiasedness of $\hat{\sigma}_j^2$ and hence of $\hat{\sigma}^2$.

(b) First note that $\{\overline{Y}_{j\cdot};\, j = 1, \ldots, J\}$ is a sequence of independent variables with expectation μ. In Exercise 4.1 the reader is invited to show that

$$\operatorname{Var}(\overline{Y}_{j\cdot}) = \tau^2 + \frac{\sigma^2}{w_{j\cdot}}.$$

A derivation similar to the one in part a—leaving the details to the reader—yields that

$$\operatorname{Var}(\overline{Y}_{j\cdot} - \overline{Y}_{\cdot\cdot}) = \frac{\sigma^2}{w_{j\cdot}}\left(1 - \frac{w_{j\cdot}}{w_{\cdot\cdot}}\right) + \tau^2\left(1 - 2\frac{w_{j\cdot}}{w_{\cdot\cdot}} + \frac{\sum_{j'} w_{j'\cdot}^2}{w_{\cdot\cdot}^2}\right), \tag{4.17}$$

where $\overline{Y}_{\cdot\cdot}$ is the weighted total in (4.2). We continue by computing

$$E\left[\sum_{j=1}^{J} w_{j\cdot} \left(\overline{Y}_{j\cdot} - \overline{Y}_{\cdot\cdot}\right)^2\right] = \sum_{j=1}^{J} w_{j\cdot} \operatorname{Var}\left(\overline{Y}_{j\cdot} - \overline{Y}_{\cdot\cdot}\right),$$

where the coefficient for σ^2 is $\sum_j (1 - w_{j\cdot}/w_{\cdot\cdot}) = J - 1$, in analogy with $n_j - 1$ in the proof of part (a). The coefficient for τ^2 is

$$\sum_j w_{j\cdot} \left(1 - 2\frac{w_{j\cdot}}{w_{\cdot\cdot}} + \frac{\sum_{j'} w_{j'\cdot}^2}{w_{\cdot\cdot}^2}\right) = w_{\cdot\cdot} - \sum_j w_{j\cdot}^2 / w_{\cdot\cdot}.$$

Consequently

$$E\left[\sum_{j=1}^{J} w_{j\cdot}(\overline{Y}_{j\cdot} - \overline{Y}_{\cdot\cdot})^2\right] = \tau^2\left(w_{\cdot\cdot} - \sum_j w_{j\cdot}^2 / w_{\cdot\cdot}\right) + (J-1)\sigma^2,$$

which, together with part (a), yields part (b). □

4.1.2 Comparison with Other Notation*

Traditional texts on credibility often do not speak of random effects, but use an abstract random risk parameter Θ_j. The task is then to estimate $\mu(\Theta_j) = E[Y_{jt}|\Theta_j]$. The correspondence to our notation is $\mu(\Theta_j) = V_j$. Since no inference is made on the parameter Θ_j itself, there is no loss in generality in our approach. On the other hand, the random effect case is contained in the traditional model as the case $\mu(\Theta_j) = \Theta_j$. Our choice of notation is based on the idea that random effects are easier to comprehend and more apt for use in multiplicative, GLM type, models. The first book to formulate credibility in terms of random effects models was Dannenburg et al. [DKG96], but their model is additive, while ours is multiplicative.

4.2 Credibility Estimators in Multiplicative Models

The classical Bühlmann-Straub model involves just a single multi-level factor (MLF). We have seen in the previous chapters that we often have a large number of rating factors, at least in the private lines. In motor insurance these might include one or more MLFs, e.g. the car model of Example 4.1 or the geographic zone of Example 4.3, alongside a number of *ordinary rating factors*, e.g. the age and gender of the policyholder, the age of the car or the annual mileage.

In this section, which is based on the article Ohlsson [Oh08], we show how to combine GLMs with credibility theory to obtain estimates of both ordinary factors and MLFs, simultaneously. For simplicity, we will mostly consider just a single MLF, but allow for an arbitrary number of ordinary factors.

Example 4.6 (Homeowner's insurance) Suppose that for each private house the insurance company has available information about *type of house* (one floor, two floors, etc.), *building area* (measured in squared meters and divided into intervals) and *geographic region* (e.g. postcode). Here *type of house* and *building area* may be used as ordinary rating factors, while *geographic region* typically is an MLF, cf. Example 4.3. A generalization of the model in (4.4) is then

$$E(Y_{ijt}|U_j) = \mu\gamma_1^i\gamma_2^i U_j,$$

where i refers to the cross-tabulation of *type of house* and *building area* (on list form, cf. the discussion at the beginning of Sect. 2.1), while j and t are as in the previous section. Here μ is the base premium, γ_1^i is the price relativity for *type of house*, γ_2^i is the one for *building area*, and U_j is a random effect for the MLF *geographic region*. To determine the premium, we would like to estimate γ_1^i and γ_2^i by GLMs and U_j by credibility theory.

As in the example, our key ratio is now denoted Y_{ijt}, where the i refers to a cell in the tariff given by the ordinary rating factors, i.e., the non-MLFs. This tariff cell is the cross-tabulation of R different rating factors, indexed on list form. If γ_r^i denotes the relativity for factor number r for policies in cell i, the multiplicative model is

$$E(Y_{ijt}|U_j) = \mu\gamma_1^i\gamma_2^i\cdots\gamma_R^i U_j,$$

generalizing (4.4). Note that μ is now the mean key ratio of the *base cell*, where $\gamma_r^i = 1$; $r = 1, \ldots, R$; and as before we assume that $E(U_j) = 1$. For the moment assuming μ and the γ_r^i's to be known, we look for a credibility estimator of U_j. For simplicity in notation we introduce

$$\gamma_i = \gamma_1^i\gamma_2^i\cdots\gamma_R^i,$$

and again $V_j = \mu U_j$, so that the multiplicative model can be rewritten

$$E(Y_{ijt}|V_j) = \gamma_i V_j. \tag{4.18}$$

If the GLM model for $Y_{jt}|V_j$ is a Tweedie model, then we have

$$\mathrm{Var}(Y_{ijt}|V_j) = \frac{\phi(\gamma_i V_j)^p}{w_{ijt}}.$$

Again let $\sigma^2 \doteq \phi E[V_j^p]$, so that

$$E[\mathrm{Var}(Y_{ijt}|V_j)] = \frac{\gamma_i^p\sigma^2}{w_{ijt}}, \tag{4.19}$$

generalizing (4.7).

We have the following extension of Assumption 4.1.

Assumption 4.2

(a) The groups, i.e., the random vectors $(V_j, Y_{1j1}, Y_{1j2}, \ldots, Y_{2j1}, Y_{2j2}, \ldots)$ are independent for $j = 1, \ldots, J$.

(b) The V_j; $j = 1, 2, \ldots, J$; are identically distributed with $E[V_j] = \mu > 0$ and $\mathrm{Var}(V_j) = \tau^2$ for some $\tau^2 > 0$.
(c) For any j, conditional on V_j, the Y_{ijt}'s are mutually independent, with mean given by (4.18) and variance satisfying (4.19).

Next we transform the observations in a way suitable for bringing back this situation to the classical Bühlmann-Straub model in Sect. 4.1;

$$\widetilde{Y}_{ijt} = \frac{Y_{ijt}}{\gamma_i}, \qquad \widetilde{w}_{ijt} = w_{ijt}\gamma_i^{2-p}. \tag{4.20}$$

Note that

$$E(\widetilde{Y}_{ijt}|V_j) = V_j,$$

and

$$E[\mathrm{Var}(\widetilde{Y}_{ijt}|V_j)] = \frac{\sigma^2}{\widetilde{w}_{ijt}}.$$

By this and Assumption 4.2, the entire Assumption 4.1 is satisfied for $\widetilde{Y}_{ijt}$ with the weights $\widetilde{w}_{ijt}$ and we get the following result from (4.9), (4.11) and (4.12).

Theorem 4.3 *Under Assumption* 4.2, *the credibility estimator of* V_j *is*

$$\widehat{V}_j = \tilde{z}_j \overline{\widetilde{Y}}_{\cdot j \cdot} + (1 - \tilde{z}_j)\mu, \tag{4.21}$$

where

$$\tilde{z}_j = \frac{\widetilde{w}_{\cdot j \cdot}}{\widetilde{w}_{\cdot j \cdot} + \sigma^2/\tau^2}, \tag{4.22}$$

and

$$\overline{\widetilde{Y}}_{\cdot j \cdot} = \frac{\sum_{i,t} \widetilde{w}_{ijt}\widetilde{Y}_{ijt}}{\widetilde{w}_{\cdot j \cdot}} = \frac{\sum_{i,t} \widetilde{w}_{ijt} Y_{ijt}/\gamma_i}{\sum_{i,t} \widetilde{w}_{ijt}}, \tag{4.23}$$

with $\widetilde{Y}_{ijt}$ *and* $\widetilde{w}_{ijt}$ *defined in* (4.20).

Consequently,

$$\widehat{U}_j = \tilde{z}_j \frac{\overline{\widetilde{Y}}_{\cdot j \cdot}}{\mu} + (1 - \tilde{z}_j), \tag{4.24}$$

and the rating for policies with MLF level j in tariff cell i is $\mu\gamma_i\widehat{U}_j$.

Remark 4.2 When Y_{ijt} is *claim frequency*, with w_{ijt} as the number of policy years, we use the Poisson GLM with $p = 1$, so that $\widetilde{w}_{ijt} = w_{ijt}\gamma_i$, a quantity that is called "normalized insurance years" in Campbell [Ca86]. Then

$$\frac{\overline{\overline{Y}}_{\cdot j \cdot}}{\mu} = \frac{\sum_{i,t} w_{ijt} Y_{ijt}}{\sum_{i,t} w_{ijt} \mu \gamma_i},$$

which is just the number of claims in group j divided by the expected number of claims in the same group: a very natural estimator of U_j.

For the case when Y_{ijt} is *claim severity* and w_{ijt} is the number of claims, the standard approach is to use a gamma GLM with $p = 2$. Here

$$\frac{\overline{\overline{Y}}_{\cdot j \cdot}}{\mu} = \frac{\sum_{i,t} w_{ijt} Y_{ijt} / (\mu \gamma_i)}{\sum_{i,t} w_{ijt}},$$

a weighted average of the observed relative deviation of Y_{ijt} from its expectation $\mu\gamma_i$—again a simple and natural estimator of U_j.

4.2.1 Estimation of Variance Parameters

We also have to estimate the variance parameters σ^2 and τ^2. Unbiased estimators can be obtained directly from the corresponding estimators in Sect. 4.1.1; the result is

$$\hat{\sigma}_j^2 = \frac{1}{n_j - 1} \sum_{i,t} \widetilde{w}_{ijt} (\widetilde{Y}_{ijt} - \overline{\overline{Y}}_{\cdot j \cdot})^2, \tag{4.25}$$

and

$$\hat{\sigma}^2 = \frac{\sum_j (n_j - 1) \hat{\sigma}_j^2}{\sum_j (n_j - 1)}. \tag{4.26}$$

Finally,

$$\hat{\tau}^2 = \frac{\sum_j \widetilde{w}_{\cdot j \cdot} (\overline{\overline{Y}}_{\cdot j \cdot} - \overline{\overline{Y}}_{\cdots})^2 - (J-1)\hat{\sigma}^2}{\widetilde{w}_{\cdot\cdot} - \sum_j \widetilde{w}_{\cdot j \cdot}^2 / \widetilde{w}_{\cdots}}, \tag{4.27}$$

where $\overline{\overline{Y}}_{\cdots}$ is the $\widetilde{w}_{\cdot j \cdot}$-weighted average of the $\overline{\overline{Y}}_{\cdot j \cdot}$'s. Note that in this case the estimators are strictly unbiased only if γ_i is known, while in practice we plug in a GLM estimate here.

4.2.2 The Backfitting Algorithm

So far, the γ_i have acted as a priori information in the sense of [BG05, Sect. 4.13], i.e., they are assumed to be known in advance, and so has μ. When we know these parameters, we may estimate $\widehat{U}_j$ from (4.24). On the other hand, when the U_j's, or rather their estimates $\widehat{U}_j$ are known, we can estimate μ and the relativities $\gamma_1^i, \ldots, \gamma_R^i$ by GLMs, and calculate $\gamma_i = \gamma_1^i \gamma_2^i \cdots \gamma_R^i$. Here we treat $\widehat{U}_j$ as a known *offset* variable, see Sect. 3.6.3. So once we know the value of U_j we can estimate μ and γ_i, and vice versa; this leads to the idea of using an iterative procedure.

We now introduce an algorithm for simultaneous, iterative rating of ordinary factors by GLMs and an MLF by credibility, in the multiplicative model $\mu\gamma_1^i \cdots \gamma_R^i U_j$.

Step 0. Initially, let $\widehat{U}_j = 1$ for all j.
Step 1. Estimate the parameters for the ordinary rating factors by a Tweedie GLM (typically Poisson or Gamma) with log-link, using $\log(\widehat{U}_j)$ as an offset-variable. This yields $\hat{\mu}$ and $\hat{\gamma}_1^i, \ldots, \hat{\gamma}_R^i$.
Step 2. Compute $\hat{\sigma}^2$ and $\hat{\tau}^2$ as in Sect. 4.2.1, using the outcome of Step 1.
Step 3. Use (4.24) to compute $\widehat{U}_j$, using the estimates from Step 1 and 2.
Step 4. Return to Step 1 with the new $\widehat{U}_j$ from Step 3.

Repeat Step 1–4 until convergence. In some cases, a relatively large number of iterations may be necessary to get good precision: it is not unusual that 100 iterations is required for a precision of 10^{-4} in the estimators. In other cases, the convergence is much faster.

We call this the *backfitting algorithm*, due to its similarity to its namesake in additive models, see Sect. 5.4.

Remark 4.3 By Sect. 3.6.3, offsetting $\log(\widehat{U}_j)$ in Step 1 is equivalent to running a GLM with new variables y_{ijt}/u_j and weights $w_{ijt}u_j^{2-p}$. Note the analogy to the variables $\widetilde{Y}_{ijt}$ and $\widetilde{w}_{ijt}$ introduced in (4.20) and used in Step 3.

Remark 4.4 Suppose that for some j we had full credibility, i.e., $z_j = 1$. Then (4.24) could be rewritten

$$\mu\widehat{U}_j = \widetilde{\overline{Y}}_{\cdot j \cdot},$$

which can be recognized as the ML equation that would result if U_j was just another ordinary rating factor in our GLM, as will be verified in Exercise 4.8.

Now, $z_j = 1$ is just the theoretical limiting case when the exposure tends to infinity. In practice we might have a large exposure w_{ijt}, by which z_j will be close to 1; then from the above derivation we expect the credibility estimate $\widehat{U}_j$ to be close to the GLM estimate. This is a nice property, with the practical implication that we do not have to worry too much about whether an U_j should be treated as a random or a fixed effect—both approaches give approximately the same estimates. Since we often have quite varying numbers of observations for different groups j, it is nice not having to single out the groups with sufficient data from the MLF.

4.2.3 Application to Car Model Classification

As an illustration, we present some results on *car model classification* in motor hull insurance, using data from the Swedish insurance group Länsförsäkringar Alliance. The *car model* (such as Renault Scénic 1.8, Volvo V70 2.4) is an important rating factor in motor insurance. As mentioned in Example 4.1, *car model* can be viewed as an MLF, with roughly 2 500 levels. In this context, *classification* means to form more or less risk-homogenous classes of car models. This can be done by estimating U_j by credibility and then grouping the cars according to their $\widehat{U}_j$-values.

There is also a large number of ordinary rating factors such as gender, age, annual mileage etc., and a specific car model j may appear in any tariff cell i constructed from these factors. The presence of both ordinary rating factors and an MLF makes the backfitting algorithm of Sect. 4.2.2 suitable for car model classification.

As usual, we make a separate analysis of *claim frequency* using a Poisson GLM with $p = 1$ and *average claim severity* using a gamma GLM with $p = 2$. Here we present results only from the claim frequency part of the study.

An application of the backfitting algorithm with all the ordinary rating factors and car model j gave the result shown by the non-filled bars (with the legend "No auxiliaries") in the histogram in Fig. 4.1, where the bins are formed by rounding the $\widehat{U}_j$'s to the nearest first decimal number. We find that almost 30% of the car models have $\widehat{U}_j$'s in the span 0.95 to 1.05, but in most cases, the car model really makes a

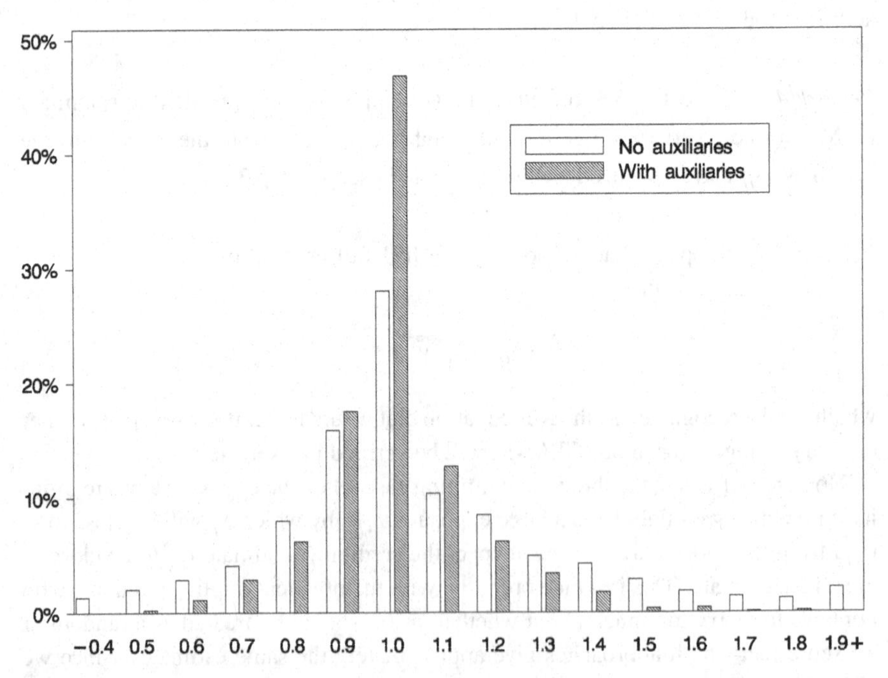

Fig. 4.1 Histogram of credibility estimators $\widehat{U}_j$ for car model, with and without auxiliary classification factors in the model

difference: the extremes differ by as much as five times ($1.9/0.4 = 4.75$) from each other.

Another approach to car model classification would be to introduce a number of ordinary rating factors that describe the car model technically, such as: weight, engine power, type of car (saloon, estate car, convertible), etc. We call these variables *auxiliaries*, since they are not intended for separate use as rating factors in the tariff, but only to help in the car model classification. Let us call this *the factor method* of car model classification; it has the advantage of not requiring any data for individual car models—a new model can be rated with the same accuracy as an old one.

It is actually possible to have the best from both worlds, by combining the credibility method and the factor method as follows. Add the auxiliaries to the ordinary factors in the GLM analysis and then again use the backfitting algorithm. The variation of the $\widehat{U}_j$'s is shown in Fig. 4.1, with the legend "With auxiliaries". Even though over 45% of the car models now get $\widehat{U}_j$'s close to 1, there is still a considerable variation among them; this shows that the factor method leaves a lot of residual variation, which can be estimated as a (residual) random effect U_j. A possible explanation is that the claim frequency is to a large extent determined by driving behavior and that different cars might attract different kinds of drivers, not only due to technical differences.

On the other hand, the introduction of the auxiliaries reduce the variation of the $\widehat{U}_j$'s. This is an indication that we get a better tariff from inclusion of auxiliaries in our credibility car model classification: a brand new or rare car model will get $\widehat{U}_j = 1$ because of lack of data, but the auxiliaries still provide some information on this car, thereby reducing the non-explained variation in claim frequency. For models with large amounts of data, on the other hand, the introduction of auxiliary variables makes little difference.

This effect is shown in some detail in Table 4.3, for a sample of car models sorted by the exposure weights $w_{\cdot j \cdot}$ (number of policy years). As expected, with large $w_{\cdot j \cdot}$, the weighted averages of the observations $\overline{\overline{Y}}_{\cdot j \cdot}$ produce reliable estimates that are given high weight in the credibility formula, and the rating of car models is hardly affected by the introduction of auxiliaries. This can be seen by comparing the classification variable in the "No auxiliaries" case, viz. $\widehat{U}_j$, to the classification variable in the "With auxiliaries" case, c_j, which is the new $\widehat{U}_j$ from this round multiplied by all the auxiliaries' relativities. Note that the auxiliaries are tied to the car model j and not to the tariff cell i.

At the other end of the table, with data from very few policy years, credibility is low and the $\overline{\overline{Y}}_{\cdot j \cdot}$'s are shaky. Here one has to rely, to a large extent, on the auxiliaries. Without them, the classification would have been much closer to 1 for these cars.

4.2.4 More than One MLF

In applications we might have more than one MLF: in motor insurance, besides car model, we may use a geographical area defined by postcode or parish, say. In

Table 4.3 Selected car models j with number of policy years $w_{\cdot j \cdot}$, weighted averages $\overline{\overline{Y}}_{\cdot j \cdot}$, credibility estimators $\widehat{U}_j$ and credibility factors $\tilde{z}_j$; without and with auxiliary factors. In the second case we have a classification variable c_j which is $\widehat{U}_j$ multiplied by the relativities for all the auxiliaries

j	$w_{\cdot j \cdot}$	No auxiliaries			With auxiliaries			
		$\overline{\overline{Y}}_{\cdot j \cdot}$	$\widehat{U}_j$	$\tilde{z}_j$	$\overline{\overline{Y}}_{\cdot j \cdot}$	$\widehat{U}_j$	$\tilde{z}_j$	c_j
1	41275	0.74	0.74	1.00	0.98	0.98	0.99	0.75
2	39626	0.58	0.58	1.00	0.89	0.89	0.99	0.59
3	39188	0.59	0.59	1.00	0.86	0.86	0.99	0.60
4	31240	0.82	0.82	1.00	0.93	0.93	0.99	0.82
5	28159	0.49	0.50	1.00	0.74	0.75	0.98	0.50
⋮	⋮	⋮	⋮	⋮	⋮	⋮	⋮	⋮
401	803	2.08	1.95	0.88	1.43	1.35	0.82	1.99
402	802	0.97	0.97	0.86	1.11	1.08	0.70	0.95
403	801	1.77	1.66	0.86	1.54	1.40	0.74	1.62
404	799	0.74	0.78	0.86	0.83	0.88	0.69	0.79
405	798	1.32	1.27	0.86	0.73	0.78	0.82	1.41
⋮	⋮	⋮	⋮	⋮	⋮	⋮	⋮	⋮
901	181	1.38	1.22	0.58	1.14	1.06	0.42	1.29
902	180	1.61	1.38	0.63	0.91	0.95	0.56	1.70
903	180	2.28	1.76	0.59	1.35	1.18	0.51	2.01
904	179	0.79	0.88	0.56	0.86	0.95	0.34	0.88
905	179	2.38	1.80	0.58	1.52	1.25	0.48	1.98
⋮	⋮	⋮	⋮	⋮	⋮	⋮	⋮	⋮
1801	7	2.39	1.07	0.05	2.05	1.03	0.03	1.22
1802	7	4.63	1.19	0.05	3.86	1.08	0.03	1.31
1803	7	0.00	0.96	0.04	0.00	0.99	0.01	0.55
1804	7	0.00	0.95	0.05	0.00	0.98	0.02	0.87
1805	7	0.00	0.94	0.06	0.00	0.98	0.02	0.58
⋮	⋮	⋮	⋮	⋮	⋮	⋮	⋮	⋮

principle, the algorithm can be extended to the case with two or more independent MLFs in a rather straight-forward fashion: in the GLM run, both variables are offset; when estimating one of the MLFs, the current estimate of the other is treated as any other rating factor.

In practice it is common, however, to rate one MLF at a time. This means that we form car model classes in one application of the algorithm, and then form geographical zones in a separate run, with the car models classes as offsets or as ordinary rating factors. A reason for this is that the final classification is often set after extensive dialogue with the subject matter experts.

4.3 Exact Credibility*

We have defined the credibility estimator of the random effect V_j as the linear function of the observations that minimizes the MSEP in (4.10). The solution, the Bühlmann-Straub estimator, was found under assumptions on the first two moments—expectation and variance—only. But suppose we know the entire probability distribution, in the form of a Tweedie EDM for the standard rating factors. Can we then improve on the credibility estimator by assuming a distribution for V_j and minimizing the MSEP *without* the requirement of linearity?

We begin the investigation of this question by showing that the conditional expectation minimizes the MSEP—a result familiar from other parts of statistics. So we are looking for the function of the observations $h(\mathbf{Y})$ that minimizes the MSEP, i.e., $E[(h(\mathbf{Y}) - V_j)^2]$, without any restrictions.

Theorem 4.4 *The function* $h(\mathbf{Y})$ *that minimizes the MSEP in* (4.10) *is given by*

$$\hat{h}(\mathbf{Y}) \doteq E(V_j|\mathbf{Y}).$$

Proof In the proof, we exclude the index j, and write $h = h(\mathbf{Y})$. First note that

$$(V - h)^2 = (V - \hat{h})^2 + (\hat{h} - h)^2 + 2(V - \hat{h})(\hat{h} - h),$$

where the last term has zero expectation, since

$$E[(V - \hat{h})(\hat{h} - h)|\mathbf{Y}] = (\hat{h} - h)(E[V|\mathbf{Y}] - \hat{h}) = 0,$$

where we have used the fact that both h and $\hat{h}$ are functions of $\mathbf{Y}$. Hence,

$$E[(V - h)^2] = E[(V - \hat{h})^2] + E[(\hat{h} - h)^2] \geq E[(V - \hat{h})^2],$$

with equality if and only if $h = \hat{h}$, i.e., when $h(\mathbf{Y}) = E(V|\mathbf{Y})$, which proves that this choice of h minimizes the MSEP. □

Theorem 4.4 does not necessarily give a practical solution to the problem of estimating V_j and U_j. But in the case when the observations follow a Tweedie EDM, we can get explicit results by assuming a so called *conjugate prior* for U_j. It turns out that the solution is then the same that we found in Sect. 4.1. From a practical point of view, this means that the extra information on probability distributions gives no new results: the same estimators are derived in a new fashion. However, it is interesting to know that—at least when we use a Tweedie model with conjugate prior—the linear credibility estimator can not be improved on.

This result is referred to as *exact credibility*, and is due to Jewell [Je74]. In the case of combining credibility with GLMs, an exact credibility result was given by Ohlsson and Johansson [OJ06].

Remark 4.5 Lee and Nelder [LN96] introduced an extension of GLMs called hierarchical generalized linear models (HGLMs), where random effects are incorporated in GLMs. By extending the likelihood concept, estimators of U_j are derived by maximizing the extended likelihood.

For Tweedie models, using conjugate distributions for U_j, the HGLM estimators are very similar to the linear credibility estimators. For a recent account of HGLMs, see [LNP06].

4.4 Hierarchical Credibility Models

In the car model classification example of Sect. 4.2.3, we might expect that cars of the same brand (Renault, Volvo, etc.) have some risk properties in common, even if they are of different models (Renault Megane, Renault Laguna, etc.). Thus one would like to be able to include *brand* as a rating factor in our models; this would be particularly helpful when a new model of a well-known brand is introduced on the market, with no insurance data available. For some brands there are lots of data (in Sweden Volvo would be an example), while for others there are very few observations; hence *brand*, like *car model*, is an MLF, for which one wants to use a credibility estimator.

Since the car models are hierarchically ordered under the brands, this is a case for using *hierarchical credibility*. As in the earlier sections, we like to take into account a priori information from the standard rating factors (age, gender, mileage, etc.) estimated by GLM and generalize the backfitting algorithm from Sect. 4.2 to the hierarchical case.

Remark 4.6 Another hierarchical MLF in motor insurance is the geographical zone, where we may have three or more levels, such as zip codes within counties, within states. For the sake of simplicity, we only consider two-level hierarchical models here.

Let Y_{ijkt} be the observed key ratio, with exposure weight w_{ijkt}, in a priori tariff cell i, for *sector* j (car brand) and *group* k (car model) within sector j. Note that while k is hierarchical under j, a particular i can be combined with any j,k and is hence *not* part of the hierarchical ordering. The full multiplicative model is now, with U_j as the random effect for sector and U_{jk} as the random effect for group within sector

$$E(Y_{ijkt}|U_j, U_{jk}) = \mu\gamma_1^i\gamma_2^i\cdots\gamma_R^i U_j U_{jk}. \tag{4.28}$$

Here we assume $E[U_j] = 1$ and $E(U_{jk}|U_j) = 1$. As before, we use an abbreviated notation for the standard rating factors, $\gamma_i = \gamma_1^i\gamma_2^i\cdots\gamma_R^i$. We need credibility estimators of both U_j and U_{jk}, but again find it easier to work with $V_j = \mu U_j$, and also with $V_{jk} = \mu U_j U_{jk} = V_j U_{jk}$, so that

$$E(Y_{ijkt}|V_j, V_{jk}) = \gamma_i V_{jk}. \tag{4.29}$$

If, given the random effects, Y follows a Tweedie GLM model, then

$$\mathrm{Var}(Y_{ijkt}|V_j, V_{jk}) = \frac{\phi(\gamma_i V_{jk})^p}{w_{ijkt}},$$

where ϕ is the dispersion parameter. This motivates the assumption

$$E[\mathrm{Var}(Y_{ijkt}|V_j, V_{jk})] = \frac{\gamma_i^p \sigma^2}{w_{ijkt}}, \tag{4.30}$$

where $\sigma^2 = \phi E[V_{jk}^p]$. We now generalize Assumption 4.2 to the hierarchical case.

Assumption 4.3

(a) The sectors, i.e., the random vectors (Y_{ijkt}, V_j, V_{jk}); $j = 1, \ldots, J$; are independent.

(b) For every j, conditional on the sector effect V_j, the groups, i.e., the random vectors (Y_{ijkt}, V_{jk}); $k = 1, \ldots, K$; are conditionally independent.

(c) All the pairs (V_j, V_{jk}); $j = 1, 2, \ldots, J$; $k = 1, 2, \ldots, K_j$; are identically distributed, with

$$E[V_j] = \mu > 0 \quad \text{and} \quad E(V_{jk}|V_j) = V_j.$$

We use the notation

$$\tau^2 \doteq \mathrm{Var}[V_j] \quad \text{and} \quad \nu^2 \doteq E[\mathrm{Var}(V_{jk}|V_j)].$$

(d) For any (j, k), conditional on (V_j, V_{jk}), the Y_{ijkt}'s are mutually independent, with mean given by (4.29) and with variance satisfying (4.30).

The hierarchical counterparts of (4.20) and (4.23) are

$$\widetilde{Y}_{ijkt} = \frac{Y_{ijkt}}{\gamma_i}, \qquad \widetilde{w}_{ijkt} = w_{ijkt}\gamma_i^{2-p}, \quad \text{and} \quad \overline{\widetilde{Y}}_{\cdot jk\cdot} = \frac{\sum_{i,t} \widetilde{w}_{ijkt} \widetilde{Y}_{ijkt}}{\sum_{i,t} \widetilde{w}_{ijkt}}. \tag{4.31}$$

The estimators of the sector and group random effects V_j and V_{jk} are given in the following two results.

Theorem 4.5 *Under Assumption* 4.3, *the credibility estimator of* V_j *is*

$$\widehat{V}_j = q_j \overline{\widetilde{Y}}^z_{\cdot j\cdot\cdot} + (1 - q_j)\mu, \tag{4.32}$$

where

$$z_{jk} = \frac{\widetilde{w}_{\cdot jk\cdot}}{\widetilde{w}_{\cdot jk\cdot} + \sigma^2/\nu^2}, \qquad \overline{\widetilde{Y}}^z_{\cdot j\cdot\cdot} = \frac{\sum_k z_{jk} \overline{\widetilde{Y}}_{\cdot jk\cdot}}{\sum_k z_{jk}} \quad \textit{and} \quad q_j = \frac{z_{j\cdot}}{z_{j\cdot} + \nu^2/\tau^2}. \tag{4.33}$$

The proof is given after the next theorem. Note that the mean $\overline{\overline{Y}}^z_{\cdot j\cdot\cdot}$ is weighted by z instead of w. If the group variance ν^2 is small then $z_{jk} \approx \nu^2 \widetilde{w}_{\cdot jk}/\sigma^2$ and so $q_j \approx \widetilde{w}_{\cdot j\cdot}/(\widetilde{w}_{\cdot j\cdot} + \sigma^2/\tau^2)$, so that q_j, as it should, is the credibility factor in a model with no groups within the sectors, cf. (4.22). If, on the other hand, $\tau^2 \approx 0$, then $q_j \approx 0$ and so $\widehat{V}_j \approx \mu$, indicating that the sector level may be omitted, as it should when not contributing to the model variance.

Theorem 4.6 *The credibility estimator of V_{jk} is*

$$\widehat{V}_{jk} = z_{jk}\overline{\overline{Y}}_{\cdot jk\cdot} + (1 - z_{jk})\widehat{V}_j, \tag{4.34}$$

where $\overline{\overline{Y}}_{\cdot jk\cdot}$, $\widehat{V}_j$ and z_{jk} are given in (4.31), (4.32) *and* (4.33), *respectively.*

By the simple relation between the U's and V's we finally note that the random effects in the multiplicative model $\mu\gamma_1^i\gamma_2^i\cdots\gamma_R^i U_j U_{jk}$ can be estimated as follows,

$$\widehat{U}_j = q_j\frac{\overline{\overline{Y}}^z_{\cdot j\cdot\cdot}}{\mu} + (1 - q_j), \qquad \widehat{U}_{jk} = z_{jk}\frac{\overline{\overline{Y}}_{\cdot jk\cdot}}{\widehat{V}_j} + (1 - z_{jk}).$$

To find the estimators of all the relativities in (4.28), i.e., $\gamma_1^i, \gamma_2^i, \ldots, \gamma_R^i, U_j$, and U_{jk}, we iterate between GLM and hierarchical credibility by an obvious extension of the backfitting algorithm from Sect. 4.2.2; we omit the details here.

Proof of Theorem 4.5 We shall bring this result back on Theorem 4.1 by showing that Assumption 4.1 is fulfilled with Y_{jt}, w_{jt} and σ^2 replaced by, respectively, $\overline{\overline{Y}}_{\cdot jk\cdot}$, z_{jk} and ν^2. By (4.29) we have $E(\overline{\overline{Y}}_{\cdot jk\cdot}|V_j, V_{jk}) = V_{jk}$, and by Assumption 4.3(c), this yields

$$E(\overline{\overline{Y}}_{\cdot jk\cdot}|V_j) = V_j, \tag{4.35}$$

which is the counterpart of (4.5). As for the variance we find, by the rule of calculating variance by conditioning and (4.30)

$$\mathrm{Var}(\overline{\overline{Y}}_{\cdot jk\cdot}|V_j) = E[\mathrm{Var}(\overline{\overline{Y}}_{\cdot jk\cdot}|V_j, V_{jk})|V_j] + \mathrm{Var}[E(\overline{\overline{Y}}_{\cdot jk\cdot}|V_j, V_{jk})|V_j]$$

$$\begin{aligned}
\Longrightarrow \quad E[\mathrm{Var}(\overline{\overline{Y}}_{\cdot jk\cdot}|V_j)] &= E[\mathrm{Var}(\overline{\overline{Y}}_{\cdot jk\cdot}|V_j, V_{jk})] + E[\mathrm{Var}(V_{jk}|V_j)] \\
&= \frac{\sum_{i,t}\widetilde{w}^2_{ijkt}E[\mathrm{Var}(Y_{ijkt}|V_j, V_{jk})]/\gamma_i^2}{\widetilde{w}^2_{\cdot jk\cdot}} + \nu^2 \\
&= \frac{\sigma^2}{\widetilde{w}_{\cdot jk\cdot}} + \nu^2.
\end{aligned}$$

By the definition of z_{jk} in (4.33), this can be rewritten as

$$E[\mathrm{Var}(\overline{\overline{Y}}_{\cdot jk\cdot}|V_j)] = \frac{\nu^2}{z_{jk}},$$

which is the counter part of (4.7) here. Note also that by Assumption 4.3(b), the $\overline{\overline{Y}}_{\cdot jk\cdot}$ are independent for different k, given V_j. Now all parts of Assumption 4.1 are fulfilled with Y_{jt}, w_{jt} and σ^2 replaced by, respectively, $\overline{\overline{Y}}_{\cdot jk\cdot}$, z_{jk} and ν^2, and we get the result directly from Theorem 4.1. □

Proof of Theorem 4.6 We shall verify that the equations of Lemma 4.1 are fulfilled. In the present setting these equations are, if we divide both sides of the second equation by γ_i and recall that $\widetilde{Y}_{ijkt} = Y_{ijkt}/\gamma_i$,

$$E(\widehat{V}_{jk}) = E(V_{jk}), \tag{4.36}$$

$$\mathrm{Cov}(\widehat{V}_{jk}, \widetilde{Y}_{ijk't}) = \mathrm{Cov}(V_{jk}, \widetilde{Y}_{ijk't}); \quad \forall i, k', t. \tag{4.37}$$

Here we have excluded the $\widetilde{Y}_{ij'k't}$-variables with $j' \neq j$, since independence makes the corresponding covariances equal to 0.

Now by Assumption 4.3(c), $E(V_{jk}) = E[V_j] = \mu$. By (4.35) we have $E(\overline{\overline{Y}}_{\cdot jk\cdot}) = \mu$, which we use in (4.32) to find $E(\widehat{V}_j) = \mu$. By inserting this in (4.34), we obtain $E(\widehat{V}_{jk}) = \mu$, proving that (4.36) is fulfilled. We now turn to the covariances.

By (4.29), Assumption 4.3(b) and (c), Lemma A.3 and (4.35),

$$\begin{aligned}
\mathrm{Cov}(V_{jk}, \widetilde{Y}_{ijk't}) &= \mathrm{Cov}[V_{jk}, E(\widetilde{Y}_{ijk't}|V_j, V_{jk}, V_{jk'})] = \mathrm{Cov}(V_{jk}, V_{jk'}) \\
&= E[\mathrm{Cov}(V_{jk}, V_{jk'})|V_j] + \mathrm{Cov}[E(V_{jk}|V_j), E(V_{jk'}|V_j)] \\
&= \begin{cases} \nu^2 + \tau^2, & \text{if } k' = k; \\ \tau^2, & \text{if } k' \neq k. \end{cases}
\end{aligned} \tag{4.38}$$

Next we turn to the left-hand side of (4.37) where

$$\mathrm{Cov}(\widehat{V}_{jk}, \widetilde{Y}_{ijk't}) = z_{jk}\,\mathrm{Cov}(\overline{\overline{Y}}_{\cdot jk\cdot}, \widetilde{Y}_{ijk't}) + (1 - z_{jk})\,\mathrm{Cov}(\widehat{V}_j, \widetilde{Y}_{ijk't}). \tag{4.39}$$

But since $\widehat{V}_j$ is a credibility estimator of V_j, we have by Lemma 4.1,

$$\mathrm{Cov}(\widehat{V}_j, \widetilde{Y}_{ijk't}) = \mathrm{Cov}(V_j, \widetilde{Y}_{ijk't}) = \mathrm{Cov}[V_j, E(\widetilde{Y}_{ijk't}|V_j)] = \mathrm{Cov}[V_j, V_j] = \tau^2. \tag{4.40}$$

In the case $k' \neq k$, by Assumption 4.3(b),

$$\begin{aligned}
\mathrm{Cov}(\overline{\overline{Y}}_{\cdot jk\cdot}, \widetilde{Y}_{ijk't}) &= E[\mathrm{Cov}(\overline{\overline{Y}}_{\cdot jk\cdot}, \widetilde{Y}_{ijk't}|V_j)] + \mathrm{Cov}[E(\overline{\overline{Y}}_{\cdot jk\cdot}|V_j), E(\widetilde{Y}_{ijk't}|V_j)] \\
&= 0 + \mathrm{Cov}[V_j, V_j] = \tau^2,
\end{aligned} \tag{4.41}$$

while in the case $k' = k$ we first note that

$$\begin{aligned}
\mathrm{Cov}(\overline{\overline{Y}}_{\cdot jk\cdot}, \widetilde{Y}_{ijkt}|V_j) &= E[\mathrm{Cov}(\overline{\overline{Y}}_{\cdot jk\cdot}, \widetilde{Y}_{ijkt}|V_j, V_{jk})|V_j] \\
&\quad + \mathrm{Cov}[E(\overline{\overline{Y}}_{\cdot jk\cdot}|V_j, V_{jk}), E(\widetilde{Y}_{ijkt}|V_j, V_{jk})|V_j] \\
&= \frac{\widetilde{w}_{ijkt}}{\widetilde{w}_{\cdot jk\cdot}} E[\mathrm{Var}(\widetilde{Y}_{ijkt}|V_j, V_{jk})|V_j] + \mathrm{Var}[V_{jk}|V_j].
\end{aligned}$$

We now proceed as in (4.41) and get, by (4.30), (4.31) and (4.33),

$$\begin{aligned}\mathrm{Cov}(\overline{\overline{Y}}_{\cdot jk\cdot}, \widetilde{Y}_{ijkt}) &= E[\mathrm{Cov}(\overline{\overline{Y}}_{\cdot jk\cdot}, \widetilde{Y}_{ijk't}|V_j, V_{jk})] + E[\mathrm{Cov}(V_{jk}, V_{jk}|V_j)] + \tau^2 \\ &= \frac{\widetilde{w}_{ijkt}}{\widetilde{w}_{\cdot jk\cdot}} E[\mathrm{Var}(\widetilde{Y}_{ijkt}|V_j, V_{jk})] + E[\mathrm{Var}(V_{jk}|V_j)] + \tau^2 \\ &= \frac{\sigma^2}{\widetilde{w}_{\cdot jk\cdot}} + \nu^2 + \tau^2 = \frac{\nu^2}{z_{jk}} + \tau^2.\end{aligned}$$

By inserting this, (4.41) and (4.40) into (4.39) we conclude

$$\mathrm{Cov}(\widehat{V}_{jk}, \widetilde{Y}_{ijkt}) = \begin{cases} z_{jk}(\frac{\nu^2}{z_{jk}} + \tau^2) + (1 - z_{jk})\tau^2 = \nu^2 + \tau^2, & \text{if } k' = k; \\ z_{jk}\tau^2 + (1 - z_{jk})\tau^2 = \tau^2, & \text{if } k' \neq k. \end{cases}$$

But this is the right-hand side of (4.38); hence, (4.37) is fulfilled and the proof is complete. □

4.4.1 Estimation of Variance Parameters

It remains to estimate the parameters σ^2, τ^2 and ν^2. Historically, only iterative so called pseudo-estimators were available in the literature for the hierarchical case, but recently direct, unbiased-type estimators were derived independently by Bühlmann and Gisler [BG05, Sect. 6.6] and Ohlsson [Oh05], for the case with just the hierarchical MLFs and no ordinary rating factors. These estimators can easily be extended to the case with a priori differences, with the following result, where T_{jk} is the number of observations i and t for the group (j, k), K_j is the number of groups in sector j and J is the number of sectors. The result is stated here without proof.

$$\hat{\sigma}^2 = \frac{1}{\sum_j \sum_k (T_{jk} - 1)} \sum_j \sum_k \sum_i \sum_t \widetilde{w}_{ijkt}(\widetilde{Y}_{ijkt} - \overline{\overline{Y}}_{\cdot jk\cdot})^2; \tag{4.42}$$

$$\hat{\nu}^2 = \frac{\sum_j \sum_k \widetilde{w}_{\cdot jk\cdot}(\overline{\overline{Y}}_{\cdot jk\cdot} - \overline{\overline{Y}}_{\cdot j\cdot\cdot})^2 - \hat{\sigma}^2 \sum_j (K_j - 1)}{\widetilde{w}_{\cdot\cdot\cdot\cdot} - \sum_j (\sum_k \widetilde{w}^2_{\cdot jk\cdot})/\widetilde{w}_{\cdot j\cdot\cdot}}; \tag{4.43}$$

$$\hat{\tau}^2 = \frac{\sum_j z_{j\cdot}(\overline{\overline{Y}}^z_{\cdot j\cdot\cdot} - \overline{\overline{Y}}^z_{\cdot\cdot\cdot\cdot})^2 - \hat{\nu}^2(J - 1)}{z_{\cdot\cdot} - \sum_j z^2_{j\cdot}/z_{\cdot\cdot}}, \tag{4.44}$$

where $\overline{\overline{Y}}_{\cdot jk\cdot}$ is given by (4.31), $\overline{\overline{Y}}^z_{\cdot j\cdot\cdot}$ by (4.33), and

$$\overline{\overline{Y}}_{\cdot j\cdot\cdot} = \frac{\sum_k \widetilde{w}_{jk}\overline{\overline{Y}}_{\cdot jk\cdot}}{\sum_k \widetilde{w}_{jk}} \quad \text{and} \quad \overline{\overline{Y}}^z_{\cdot\cdot\cdot\cdot} = \frac{\sum_j z_{j\cdot}\overline{\overline{Y}}^z_{\cdot j\cdot\cdot}}{\sum_j z_{j\cdot}}.$$

4.4.2 Car Model Classification, the Hierarchical Case

As discussed above, the *car brand* might contain valuable information that is not contained in the auxiliary (technical) data on the car; this suggest the application of a hierarchical credibility model with the MLF *car model* hierarchically ordered under the MLF *car brand*. In total the multiplicative model then includes: the ordinary rating factors, the auxiliary factors based on technical properties of the car models, the *car brand* random effect, and finally the *car model* random effect. We summarize this in Table 4.4, using A to denote the number of auxiliaries.

Note that even though *car brand* is a random effect in the hierarchical credibility model, from a practical point of view it serves the same purpose as the auxiliaries, as explained in Sect. 4.2.3. The multiplicative model is now

$$E(Y_{ijkt}|U_j, U_{jk}) = \mu\gamma_1^i \cdots \gamma_R^i \gamma_{R+1}^{jk} \cdots \gamma_{R+A}^{jk} U_j U_{jk}.$$

As already mentioned, this model can be estimated by a hierarchical extension of the backfitting algorithm of Sect. 4.2.2. We will not report all the extensive output from such an analysis, but in Fig. 4.2 we display $\widehat{U}_j$ for the 88 car brands used in the analysis of hull insurance at Länsförsäkringar Alliance. We see that there are substantial differences between many of the car brands, but not as large as for car models in Fig. 4.1. We conclude that it is worthwhile to include *car brand* in our multiplicative model; this is especially important for the classification of car models on which we have none or few data.

After separately analyzing claim frequency and claim severity by the above method, we multiply the auxiliary factors, the car brand factor $\widehat{U}_j$ and the car model factor $\widehat{U}_{jk}$, from each of these two analyzes and get

$$c_{jk} = \gamma_{R+1}^{jk} \cdots \gamma_{R+A}^{jk} \widehat{U}_j \widehat{U}_{jk}.$$

The c_{jk}'s for claim frequency and severity are then multiplied, giving a classification variable for the pure premium. The classification is finalized by calibrating this variable to maintain the total premium level, as is demonstrated in Sect. 3.4 of Sundt [Su87], and then forming classes from the calibrated variable.

This concludes our rather extensive treatment of the problem of car model classification. The combination of GLMs with (hierarchical) credibility in the backfitting algorithm is an efficient tool here and in other situations, such as the problem of

Table 4.4 Variables in car model classification with hierarchical credibility and GLM

Rating factors	Example	Type	Notation
Ordinary rating factors	Age of driver	Fixed effect	$\gamma_1^i, \gamma_2^i, \ldots, \gamma_R^i$
Auxiliary factors	Car weight	Fixed effect	$\gamma_{R+1}^{jk}, \gamma_{R+2}^{jk}, \ldots, \gamma_{R+A}^{jk}$
Car brand	Volvo	Random effect	U_j
Car model	Volvo V70 2.4	Random effect	U_{jk}

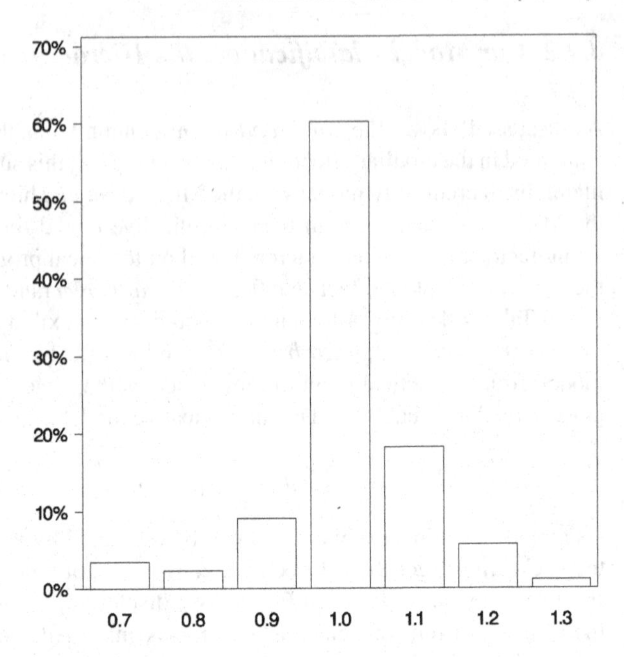

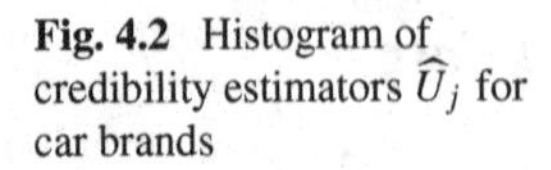
Fig. 4.2 Histogram of credibility estimators $\widehat{U}_j$ for car brands

constructing geographic zones. In practice, the statistical analysis is often supplemented by expert judgement, and the result is not exactly the relativities estimated by backfitting; hence, we may want to offset the finally chosen variable in any further analysis of the ordinary rating factors.

4.5 Case Study: Bus Insurance

Let us investigate how the (non-hierarchical) backfitting algorithm in Sect. 4.2.2 works for bus insurance in the commercial lines. Transportation companies own one or more buses which are insured for a shorter or longer period. Our aim here is to estimate the individual claim frequency of 670 companies that were policyholders at the former Swedish insurance company Wasa during the years 1990–1998. The data set `busscase.txt`, available at www.math.su.se/GLMbook, contains the following variables (with Swedish acronyms):

- *ZON*: geographic subdivision of Sweden into seven zones, based on parishes and numbered 1–7. The zones are the same as for moped insurance, Example 1.1, see Table 1.1.
- *BUSSALD*: the age class of the bus, in the span 0–4.
- *KUNDNR*: an ID number for the company, re-coded here for confidentiality reasons.
- *ANTAVT*: number of observations for the company in a given tariff cell based on zone and age class. There may be more than one observation per bus, since each renewal is counted as a new observation.

- *DUR*: duration measured in days and aggregated over all observations in the tariff cell.
- *ANTSKAD*: the corresponding number of claims.
- *SKADKOST*: the corresponding total claim cost.

The premium is based on the unit *bus year*; the premium for a bus company is the sum of the premiums for all its buses, in all tariff cells.

Problem 1: Compute relativities for the claim frequency in a multiplicative tariff based on ZON and BUSSALD, without reference to the fact that the buses belong to different companies. For simplicity, we accept the levels of the rating factors as they are and let $\phi = 1$.
Hint: There is a company with some obvious errors in the data. Exclude that company from the analysis. (In practice we would try to get the correct value from some source.)

Problem 2: Next take into account that companies may be different by using KUNDNR as an MLF. Use the backfitting algorithm in Sect. 4.2.2 with $p = 1$. Do you notice any change in the relativities for ZON or BUSSALD? Investigate how large the $\hat{u}_j$'s are by making a graph.
Hint: In the case $p = 1$, we can write $\mu_i = \mu\gamma_i$; in this notation

$$\tilde{z}_j = \frac{\sum_{i,t} w_{ijt}\mu_i}{\mu\sigma^2/\tau^2 + \sum_{i,t} w_{ijt}\mu_i} \quad \text{and} \quad \frac{\tilde{\overline{Y}}_{\cdot j\cdot}}{\mu} = \frac{\sum_{i,t} w_{ijt} Y_{ijt}}{\sum_{i,t} w_{ijt}\mu_i}.$$

Here we need to estimate $\mu\sigma^2/\tau^2$, but this value can not be computed accurately with the aggregated data that is provided. For simplicity, we use the estimate $\mu\sigma^2/\tau^2 = 2.81$ without computation; note that in practice, this value would of course change in each iteration.

Problem 3: Test what would happen if $\mu\sigma^2/\tau^2 = 28.1$, instead of 2.81.

In reality the next step would be to estimate the average claim severity, either in a standard GLM on just ZON and BUSSALD, or with the same technique as above. Here we expect lower credibility than for claim frequency, so we might just stick to the ordinary rating factors in this case.

Exercises

4.1 (Section 4.1)

(a) Use (4.8) and Lemma A.2 to show that

$$\text{Var}\left(\overline{Y}_{j\cdot}\right) = \tau^2 + \frac{\sigma^2}{w_{j\cdot}}.$$

(b) Show that the credibility factor in (4.12) can be interpreted as the percentage of the variance of the observations that emanates from the variance between groups, i.e., $\tau^2 = \text{Var}(V_j)$.

4.2 (Section 4.1) Suppose that in a commercial insurance line of business, we have computed the pure premium for a company j, based on data from n years: $t = 1, 2, \ldots, n$, by using $\widehat{V}_j$ in Theorem 4.1, for the moment denoted as $\widehat{V}_j^n$. Then a year passes and we get new data $(w_{j,n+1}, Y_{j,n+1})$ and want to compute the new credibility estimator with as little effort as possible. Suppose that both σ^2/τ^2 and μ are unchanged by the addition of the new data. Show that then the new credibility estimator $\widehat{V}_j^{n+1}$ can be written

$$\widehat{V}_j^{n+1} = cY_{j,n+1} + (1 - c)\widehat{V}_j^n,$$

and determine the value of c.

4.3 (Section 4.1) Repeat the calculations in Example 4.5, in a spreadsheet. Then vary the Y-values and note the impact on the credibility factors. What if we have a single very large Y-value? Also try deleting some observations of a company and see how its credibility decreases.

4.4 (Section 4.1) With the data in Example 4.5, compute the two variance components in Exercise 4.1 for each company.

4.5 (Section 4.1) Return once more to Example 4.5. Show that the credibility factors in (4.12) are unaltered by changing unit of measure for the weights w (e.g. by using number of employees instead of 10th of employees). *Hint*: Introduce new weights $w* = cw$ for some scaling factor c and see how σ^2 and τ^2 or $\hat{\sigma}^2$ and $\hat{\tau}^2$ change.

4.6 (Section 4.2) We obtained Theorem 4.3 by bringing it back on Theorem 4.1. Alternatively, we could have derived Theorem 4.3 directly from Lemma 4.1. If so, how could Theorem 4.1 be obtained as a special case of Theorem 4.3?

4.7 (Section 4.2) In Example 4.5 and Exercise 4.3 we computed credibility estimators for five companies. Now, suppose that we already had a tariff with three classes, in which we had obtained $\hat{\mu} = 481.3$, $\hat{\gamma}_1 = 0.6$, $\hat{\gamma}_2 = 1.0$ and $\hat{\gamma}_3 = 1.4$. A company may well belong to one class for one year and then to another for the next year, due perhaps to a change in composition of its work force.

Say that for all years, company 1 belonged to class 2 and company 2 belonged to class 1. Company 3 was in class 2 except for the years 3 and 6 when it was in class 3. Company 4 belonged to class 2 until year 10 when it changed to class 3 and, finally, company 5 belonged to class 3 for the first 7 years and then to class 2 for year 8–9.

Recalculate the credibility estimates with this extra information. In this case, we do not know the value of p, since we have no information on how the tariff was estimated. Try different values of p and see how it effects the estimates and in particular the credibility factors, say $p = 1.0$; 1.5 and 2.0.

4.8 (Section 4.2) Prove the statement in Remark 4.4 that $\widehat{U}_j = \widetilde{\overline{Y}}_{\cdot j \cdot}/\mu$, defines the maximum likelihood estimating equations for U_j in the corresponding GLM. Note that the latter equations in the present notation would be, from (2.30),

$$\sum_i \sum_t \frac{w_{ijt}}{\mu_{ij}^{p-1}} y_{ijt} = \sum_i \sum_t \frac{w_{ijt}}{\mu_{ij}^{p-1}} \mu_{ij}, \tag{4.45}$$

where $\mu_{ij} = \mu \gamma_i u_j$.

Chapter 5
Generalized Additive Models

Almost every non-life insurance pricing problem involves continuous rating variables, like the age of the policyholder or the engine power of an insured vehicle. By far, the most common approach in practice is to group the possible values of the variable into intervals, treating values in the same interval as identical. This method has the advantage of being simple and often works well enough. However, an obvious disadvantage is that the premium for two policies with different but close values for the rating variable may have substantially different premiums if the values happen to belong to different intervals. Also, for a variable like age, it is unsatisfactory if the premium is more or less constant for a number of years and then suddenly goes up or down when the end of the interval is reached. Finally, finding a good subdivision into intervals is often time consuming and tedious. The intervals must be large enough to achieve good precision of the estimates, but at the same time have to be small if the effect of the variable varies much. Sometimes both of these requirements cannot be met.

Consider for instance the data in Fig. 5.1, which gives a clear impression of an underlying continuous curve. The interval subdivision method corresponds to modeling the effect by means of a step function. This is only an approximation to the underlying continuous curve, which is not satisfactory unless we have a large number of steps. One may try to smooth the estimated relativities afterwards, but the statistical efficiency of this technique is questionable. If we really believe that the underlying effect varies continuously, then what we would like to have is a model where this is incorporated as such. This would give us a better description of the underlying risk, and hence a more competitive tariff. A simple way to achieve this is to use polynomial regression, which was briefly discussed in Sect. 3.6.4. Although polynomial regression is a step forward from the interval subdivision method and works well in many situations, it turns out that there are some problems also with this technique. To begin with, increasing the fit proceeds in discrete steps, by increasing the degree of the polynomial, and the shape of the curve can change dramatically in a single step. A more serious problem is that high degree polynomials are unruly and sensitive, causing numerical problems. Furthermore, it has been observed that the value of the polynomial at a certain point may be heavily influenced by observations pertaining to values in a completely different range.

E. Ohlsson, B. Johansson, *Non-Life Insurance Pricing with Generalized Linear Models*, EAA Lecture Notes,
DOI 10.1007/978-3-642-10791-7_5, © Springer-Verlag Berlin Heidelberg 2010

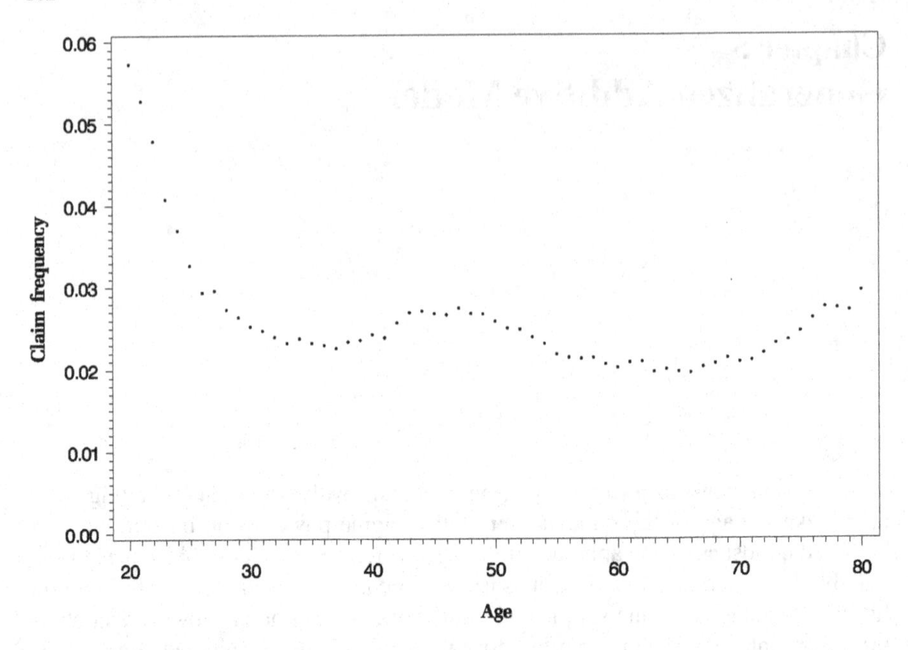

Fig. 5.1 Claim frequencies for motor TPL versus policyholder age

In the 1980's the problem of how to analyze the effect of continuous variables was the object of much research and led to an extension of GLMs known as *generalized additive models* (GAMs). There exist several possible approaches to analyzing GAMs; we shall only consider the well-known approach that goes under the name of *smoothing splines*. It utilizes results from the theory of splines, i.e. piecewise polynomials, which have their origin in the field of numerical analysis. To facilitate the reading of this chapter, we have placed some technical details on splines in Appendix B, focusing on ideas and practical issues in the main text.

Implementing the methods described in this chapter is numerically more involved than has been the case up to now. When you are applying the methods in practice, you are likely to use some kind of software for GAMs, and therefore we focus mostly on explaining the basic concepts. The reader who wants to go deeper into the numerics is referred to Wood [Wo06]. This reference also contains additional topics, such as the construction of confidence regions and an overview over some other approaches to GAMs.

There is an interesting connection between GAMs and GLMs with random effects, which opens up the possibility of using estimation methods from random effect models. See for instance Lee et al. [LNP06, Sect. 8.2.3].

5.1 Penalized Deviances

The starting point is that we have a set of data on list form, like in Table 2.2, with a set of potential *rating variables* $x_{i1}, \ldots, x_{iJ}$, where x_{ij} denotes the value of variable

j for observation i. The variables may be either categorical or continuous, but so as not to complicate the situation more than necessary, we do not include multilevel factors here. In Sect. 2.2 we discussed how to specify a model for the means, arriving at (2.19):

$$\eta_i = \sum_{j=1}^{r} \beta_j x'_{ij} \quad i = 1, 2, \ldots, n. \tag{5.1}$$

Here we have written x'_{ij} instead of x_{ij} to emphasize that the covariates appearing in η_i are not the original variables but some transformation of them.

Supposing at first that we have no interactions, the model may be written as

$$\eta_i = \beta_0 + \sum_{k_1=1}^{K_1} \beta_{1k_1} \phi_{1k_1}(x_{i1}) + \cdots + \sum_{k_J=1}^{K_J} \beta_{Jk_J} \phi_{Jk_J}(x_{iJ}). \tag{5.2}$$

Note that here the β's are renumbered as compared to (5.1); this notation makes it clear to which variable a certain β parameter is associated. If variable j is categorical, assuming a limited set of values $\{z_1, \ldots, z_{K_j}\}$, then $\phi_{jk}(x_{ij})$ assumes the value 1 or 0 according to whether x_{ij} equals z_k or not. If variable j is continuous and we use a subdivision into K_j intervals, then $\phi_{jk}(x_{ij}) = 1$ if x_{ij} belongs to interval k, otherwise $\phi_{jk}(x_{ij}) = 0$. If we are using polynomial regression for variable j, we have $\phi_{jk}(x) = x^k$.

Remark 5.1 To make the distinction clear we use the term *rating variable* for the original covariate, and the term *rating factor* for the transformed covariate that appears in η_i.

Hastie and Tibshirani introduced the concept *generalized additive models*, see [HT90], where instead of (5.2) one assumes only that

$$\eta_i = \beta_0 + f_1(x_{i1}) + \cdots + f_J(x_{iJ}),$$

for some functions f_j. Thus the additive effect of each variable is retained, whereas one permits a more general model of how the mean depends on the particular value of a variable. Of course, some of the functions may be of the traditional form as in (5.2). One may also include interactions by means of functions f_j of several variables, but we shall postpone this case until Sect. 5.6.

Assume to begin with that we model all the variables except the first one in the traditional way; the model then looks like

$$\eta_i = \sum_{j=1}^{r} \beta_j x'_{ij} + f(x_{i1}). \tag{5.3}$$

Insurance data may consist of several million observations, but due to rounding a continuous variable usually assumes a much smaller number of values. For instance, the age of the policyholder is usually registered as an integer, and so less than a hundred different values occur. We let $z_1, \ldots, z_m$ denote the possible values of x_{i1}, sorted in ascending order. As already mentioned, the number of values m is typically much smaller than the number of observations n.

As a measure of the goodness of fit of some estimated means $\hat{\mu} = \{\hat{\mu}_i\}$ to the data, we use the deviance $D(y, \hat{\mu})$. Recall the discussion in Sect. 3.5, where it was argued that deviances, like those in (3.4), may be defined starting from a variance assumption, without reference to a likelihood. For simplicity, we shall refer to the cases $v(\mu) = 1$, $v(\mu) = \mu$ and $v(\mu) = \mu^2$ as the normal, Poisson and gamma cases, respectively, but it is to be understood that we do not assume a specific distribution, but rather use the approach of Sect. 3.5.

Our goal is to find the function f in (5.3) that in some sense gives the best description of the effect of the continuous variable. So far we have made no assumptions concerning $f(\cdot)$ and obviously two functions taking the same values at the points $z_1, \ldots, z_m$ will produce the same deviance. To begin with, modeling the effect of a continuous variable, it is reasonable to assume that $f(\cdot)$ is continuous, but in fact we expect a little more than that, since arbitrary continuous functions may still behave in an extremely irregular way. We shall therefore demand the function to be smooth, which we take to mean that it is twice continuously differentiable. Still, this is of course not enough to determine which function to use, since there are infinitely many smooth functions assuming the same values at the points $z_1, \ldots, z_m$. To proceed, we realize that apart from being smooth, we also expect the function $f(\cdot)$ not to vary wildly. We thus need a measure of the variability of $f(\cdot)$. There are several possible definitions, but an intuitively appealing one is the integrated squared second derivative, $\int_a^b (f''(x))^2\,dx$, where $[a, b]$ is some interval with $a \leq z_1$, $z_m \leq b$. For a function with much variation the second derivative will change a lot, resulting in a high integrated squared value. A function without variability in this sense will have $\int_a^b (f''(x))^2\,dx = 0$, so that the second derivative is identically zero, and the function must be a straight line. The basic idea is now to add a penalty term to the deviance and find a function f that minimizes

$$\Delta(f) = D(y, \mu) + \lambda \int_a^b (f''(x))^2\,dx, \tag{5.4}$$

where $D(y, \mu)$ depends on f through (5.3) and the relation $\eta_i = g(\mu_i)$. The parameter λ strikes a balance between good fit to the data, measured by the deviance, and variability of the function f.

Still it is not at all clear how to proceed, since all we have assumed so far is that f belongs to the set of twice continuously differentiable functions. However, in the next section we will show that it suffices to consider a subset of this general class, the so called *cubic splines*.

5.2 Cubic Splines

To construct a spline function, you take a number of polynomials, defined on a connected set of disjoint intervals, and tie the polynomials together at the points where the intervals adjoin, in such a way that a certain number of derivatives at these points match. The concept of splines can be defined in a general way, see e.g. de Boor [dBo01], but we shall only use a limited set of splines.

The simplest kind of spline consists of a set of linear functions tied together. More precisely, assume that we have a set of points $u_1, \ldots, u_m$ with $u_1 < \cdots < u_m$ and that we define the linear functions $p_k(x) = a_k + b_k x$, $k = 1, \ldots, m-1$ in such a way that the endpoints match, i.e. $p_{k-1}(u_k) = p_k(u_k)$, $k = 2, \ldots, m-1$. We then define the function $s(x)$ on the interval $[u_1, u_m]$ by

$$s(x) = p_k(x), \quad u_k \le x \le u_{k+1};\ k = 1, \ldots, m-1. \tag{5.5}$$

The function $s(x)$ is continuous, and linear between any two points u_k and u_{k+1}. We shall call this a *linear spline*.

To get a curve that is continuous and also continuously differentiable, one may instead tie together quadratic functions. Again, define $s(x)$ as in (5.5), but now using $p_k(x) = a_k + b_k x + c_k x^2$, where the parameters a_k, b_k and c_k have been chosen in such a way that $p_{k-1}(u_k) = p_k(u_k)$ and $p'_{k-1}(u_k) = p'_k(u_k)$ for $k = 2, \ldots, m-1$. In this case, we call $s(x)$ a *quadratic spline*.

Proceeding in the same manner, we can define a spline that is twice continuously differentiable by using $p_k(x) = a_k + b_k x + c_k x^2 + d_k x^3$ and in addition to the conditions for the quadratic spline add $p''_{k-1}(u_k) = p''_k(u_k)$ for $k = 2, \ldots, m-1$. We call this kind of object a *cubic spline*. We shall want to be able to extend a cubic spline to an interval $[a, b]$ containing $[u_1, u_m]$. In particular, we shall encounter cubic splines s that are linear outside (u_1, u_m), i.e. $s''(x) = 0$ for $x \in [a, u_1]$ and $x \in [u_m, b]$. Such cubic splines are called *natural*.

More generally, a function on the interval $[u_1, u_m]$ is called a spline of order j if it is $j-1$ times continuously differentiable and on each interval $[u_k, u_{k+1}]$ it is a polynomial of order j.

The points $u_1, \ldots, u_m$ are called the *knots* of the spline and $u_2, \ldots, u_{m-1}$, where the polynomial are tied together, are called the *internal knots*.

To illustrate the effect on the smoothness of the curve when increasing the order of the spline, in Fig. 5.2 a step function is shown together with a linear, a quadratic and a cubic spline, all with the same set of knots.

The reason that cubic splines play a fundamental role in connection with penalized deviances is contained in the following theorem, proved in Appendix B.1.

Theorem 5.1 *For any points $u_1 < \cdots < u_m$ and real numbers $y_1, \ldots, y_m$, there exists a unique natural cubic spline s, such that $s(u_k) = y_k$, $k = 1, \ldots, m$. Furthermore, if f is any twice continuously differentiable function such that $f(u_k) = y_k$, $k = 1, \ldots, m$, then for any $a \le u_1$ and $b \ge u_m$,*

$$\int_a^b (s''(x))^2\, dx \le \int_a^b (f''(x))^2\, dx. \tag{5.6}$$

Consider now the penalized deviance (5.4) for any twice continuously differentiable function f. The deviance part of (5.4) is uniquely determined by the values

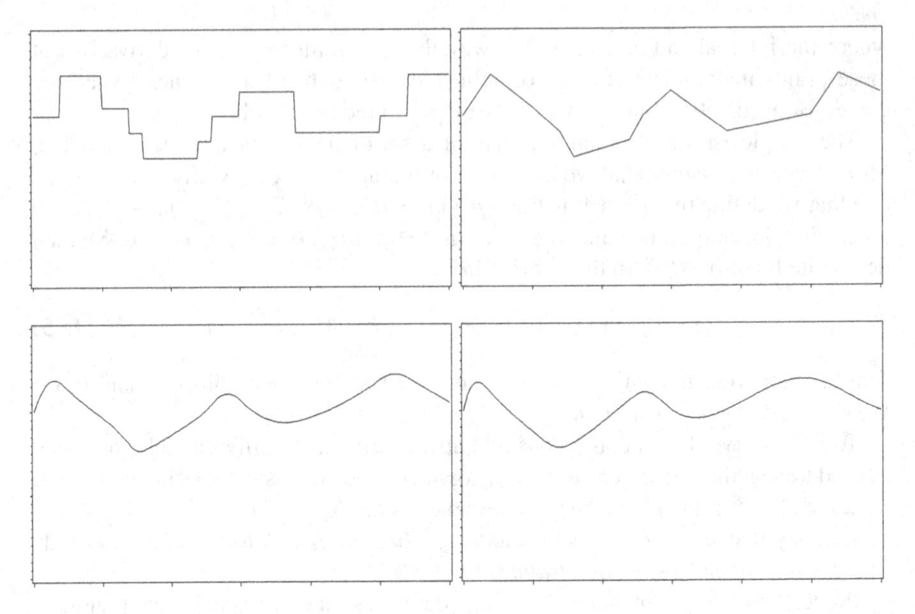

Fig. 5.2 Step function (*top left*) together with linear (*top right*), quadratic (*bottom left*) and cubic splines (*bottom right*)

$f(z_k), k = 1, \ldots, m$. By Theorem 5.1, there is a unique natural cubic spline s, such that $s(z_k) = f(z_k), k = 1, \ldots, m$, and so the deviances corresponding to f and s are identical. Thus, due to (5.6), $\Delta(s) \leq \Delta(f)$.

Splines have been much used for *interpolation*, where one is looking for a smooth function passing through a fixed set of points. Theorem 5.1 states that among all interpolating twice continuously differentiable functions, the natural cubic spline minimizes the integrated squared second derivative. Recall that there also always exists a unique interpolating polynomial of degree $m - 1$. In Fig. 5.3 the interpolating natural cubic spline is compared to the interpolating polynomial for a set of points. The figure clearly shows that the spline is a much better choice for interpolation.

Theorem 5.1 makes it possible to take a big step forward in connection with our problem of minimizing the penalized deviance (5.4). As concluded above, it implies that when looking for a minimizing function among all twice continuously differentiable functions, it suffices to consider the set of natural cubic splines. This means that we may reduce the search to a set of functions that can be characterized by means of a finite set of parameters, and so we have a well-defined minimization problem. There are several possible ways of parametrizing the set of cubic splines. We shall use the fact that a spline may be written as a linear combination of simple basis splines called *B-splines*. This is the most common approach, due to its superior numerical properties.

The definition and basic properties of B-splines are presented in Appendix B.2. Here we shall be content with the graphical impression of Fig. 5.4, which shows some cubic B-splines. The linear space of splines of order j for a set of knots $u_1, \ldots, u_m$ has dimension $m + j - 1$, which is also the number of B-splines for

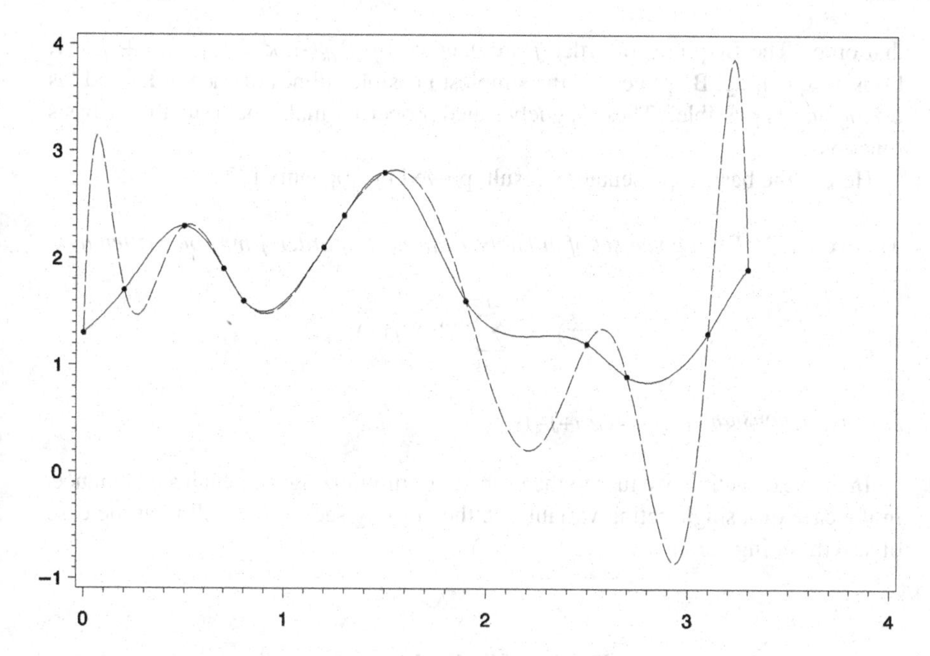

Fig. 5.3 Interpolating natural cubic spline (*solid*) compared to interpolating polynomial

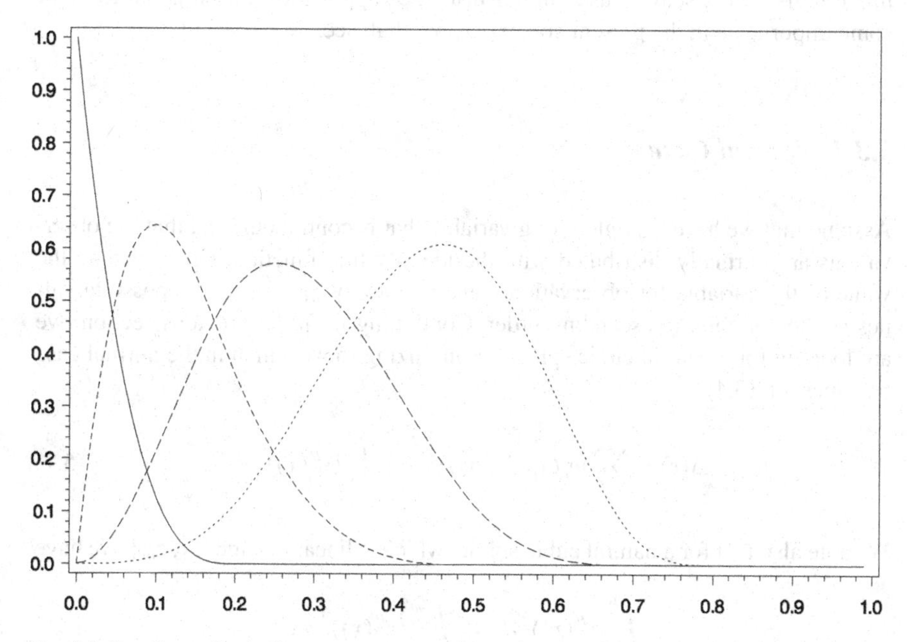

Fig. 5.4 B-splines of order 3

that order. The B-splines of order j are denoted by $B_{jk}(\cdot)$, $k = 1, \ldots, m + j - 1$. Loosely speaking, B-splines are the simplest possible splines of each order and 'as orthogonal as possible'. These vaguely stated properties make them suitable as basis functions.

Here is the basic representation result, proved in Appendix B.2.

Theorem 5.2 *For a given set of m knots, a spline s of order j may be written as*

$$s(x) = \sum_{k=1}^{m+j-1} \beta_k B_{jk}(x),$$

for unique constants $\beta_1, \ldots, \beta_{m+j-1}$.

In the next section, we turn to the problem of minimizing the penalized deviance, in the case of a single rating variable. In the ensuing section we will treat the case of several rating variables.

5.3 Estimation—One Rating Variable

Before studying our basic models for claim frequency and claim severity, we shall begin with the classic case of normally distributed observations. Although the normal distribution is seldom used in our insurance applications, it turns out to be of some importance in the present context, as we shall see.

5.3.1 Normal Case

Assume that we have a single rating variable that is continuous, and that the observations are normally distributed with the identity link function. Let x_i denote the value of the variable for observation i and denote by $z_1, \ldots, z_m$ the possible values of the variable, in ascending order. Continuing from the previous section, we are looking for a natural cubic spline s minimizing (5.4), which in the normal case becomes, cf. (3.4),

$$\Delta(s) = \sum_i w_i (y_i - s(x_i))^2 + \lambda \int_a^b (s''(x))^2 \, dx. \tag{5.7}$$

We note also that for a natural cubic spline, which is linear outside $[z_1, z_m]$, we have

$$\int_a^b (s''(x))^2 \, dx = \int_{z_1}^{z_m} (s''(x))^2 \, dx.$$

Therefore, only $s(x)$, $z_1 \le x \le z_m$ is used in the penalized deviance. On the interval $[z_1, z_m]$ we know from Theorem 5.2 that $s(x)$ may be written as

$$s(x) = \sum_{j=1}^{m+2} \beta_j B_j(x),$$

where $B_1(x), \ldots, B_{m+2}(x)$ are the cubic B-splines with knots $z_1, \ldots, z_m$. Using this, the penalized deviance may be considered as a function of the parameters $\beta_1, \ldots, \beta_{m+2}$, and becomes

$$\Delta(\boldsymbol{\beta}) = \sum_i w_i \left(y_i - \sum_{j=1}^{m+2} \beta_j B_j(x_i) \right)^2 + \lambda \sum_{j=1}^{m+2} \sum_{k=1}^{m+2} \beta_j \beta_k \Omega_{jk},$$

where

$$\Omega_{jk} = \int_{z_1}^{z_m} B_j''(x) B_k''(x)\, dx.$$

The numbers Ω_{jk} may be computed using the basic properties of B-splines. The details are given in Appendix B.2.

To find the minimizing $\beta_1, \ldots, \beta_{m+2}$, we calculate the partial derivatives:

$$\frac{\partial \Delta}{\partial \beta_\ell} = -2 \sum_i w_i \left(y_i - \sum_{j=1}^{m+2} \beta_j B_j(x_i) \right) B_\ell(x_i) + 2\lambda \sum_{j=1}^{m+2} \beta_j \Omega_{j\ell}.$$

Letting I_k denote the set of i for which $x_i = z_k$, we get

$$\begin{aligned}
-2 \sum_i & w_i \left(y_i - \sum_{j=1}^{m+2} \beta_j B_j(x_i) \right) B_\ell(x_i) \\
&= -2 \sum_{k=1}^{m} \sum_{i \in I_k} w_i \left(y_i - \sum_{j=1}^{m+2} \beta_j B_j(z_k) \right) B_\ell(z_k) \\
&= -2 \sum_{k=1}^{m} \tilde{w}_k \left(\tilde{y}_k - \sum_{j=1}^{m+2} \beta_j B_j(z_k) \right) B_\ell(z_k),
\end{aligned}$$

where

$$\tilde{w}_k = \sum_{i \in I_k} w_i, \qquad \tilde{y}_k = \frac{1}{\tilde{w}_k} \sum_{i \in I_k} w_i y_i.$$

Setting the partial derivatives equal to zero, we obtain the equations

$$\sum_{k=1}^{m}\sum_{j=1}^{m+2}\tilde{w}_k\beta_j B_j(z_k)B_\ell(z_k)+\lambda\sum_{j=1}^{m+2}\beta_j\Omega_{j\ell}=\sum_{k=1}^{m}\tilde{w}_k\tilde{y}_k B_\ell(z_k);$$
$$\ell=1,\ldots,m+2. \tag{5.8}$$

Introduce the $m\times(m+2)$ matrix $\mathbf{B}$ by

$$\mathbf{B}=\begin{pmatrix} B_1(z_1) & B_2(z_1) & \cdots & B_{m+2}(z_1)\\ B_1(z_2) & B_2(z_2) & \cdots & B_{m+2}(z_2)\\ \vdots & \vdots & \ddots & \vdots\\ B_1(z_m) & B_2(z_m) & \cdots & B_{m+2}(z_m)\end{pmatrix}.$$

Let $\mathbf{W}$ denote the $m\times m$ diagonal matrix with $\tilde{w}_j$ on the main diagonal and let $\boldsymbol{\Omega}$ denote the symmetric $(m+2)\times(m+2)$ matrix with elements Ω_{jk}. Furthermore, let $\boldsymbol{\beta}$ and $\mathbf{y}$ denote the column vectors with elements β_j and $\tilde{y}_k$, respectively. The equation system (5.8) may then be written on matrix form as

$$\left(\mathbf{B}'\mathbf{W}\mathbf{B}+\lambda\boldsymbol{\Omega}\right)\boldsymbol{\beta}=\mathbf{B}'\mathbf{W}\mathbf{y}. \tag{5.9}$$

Due to the way the B-splines are defined, the matrices $\mathbf{B}'\mathbf{W}\mathbf{B}$ and $\boldsymbol{\Omega}$ are banded, which simplifies the numerical solution of the linear equation system.

5.3.2 Poisson Case

We now move on to the Poisson case, with $g(\mu)=\log\mu$. We recall from (3.4) that the deviance is

$$2\sum_i w_i(y_i\log y_i - y_i\log\mu_i - y_i+\mu_i).$$

Since we have a single continuous rating variable, $\mu_i=\exp\{s(x_i)\}$, where s is again a natural cubic spline. The penalized deviance thus becomes

$$\Delta(s)=2\sum_i w_i(y_i\log y_i - y_i s(x_i) - y_i+\exp\{s(x_i)\})+\lambda\int_a^b (s''(x))^2\,dx.$$

The problem reduces to finding $\beta_1,\ldots,\beta_{m+2}$ minimizing

$$\Delta(\boldsymbol{\beta})=2\sum_i w_i\left(y_i\log y_i - y_i\sum_{j=1}^{m+2}\beta_j B_j(x_i)-y_i+\exp\left\{\sum_{j=1}^{m+2}\beta_j B_j(x_i)\right\}\right)$$
$$+\lambda\sum_{j=1}^{m+2}\sum_{k=1}^{m+2}\beta_j\beta_k\Omega_{jk}.$$

Differentiating with respect to the β_ℓ's and setting the derivatives equal to zero, we arrive at the equations

$$-\sum_i w_i y_i B_\ell(x_i) + \sum_i w_i \gamma(x_i) B_\ell(x_i) + \lambda \sum_{j=1}^{m+2} \beta_j \Omega_{j\ell} = 0;$$
$$\ell = 1, \ldots, m+2, \tag{5.10}$$

where we have introduced the notation

$$\gamma(x) = \exp\left\{ \sum_{j=1}^{m+2} \beta_j B_j(x) \right\}. \tag{5.11}$$

Using $\tilde{w}_k$ and $\tilde{y}_k$ as in the normal case, the equations become

$$-\sum_{k=1}^{m} \tilde{w}_k \tilde{y}_k B_\ell(z_k) + \sum_{k=1}^{m} \tilde{w}_k \gamma(z_k) B_\ell(z_k) + \lambda \sum_{j=1}^{m+2} \beta_j \Omega_{j\ell} = 0;$$
$$\ell = 1, \ldots, m+2. \tag{5.12}$$

Since $\gamma(z_k)$ depends in a nonlinear way on $\beta_1, \ldots, \beta_{m+2}$, we apply the Newton-Raphson method. Put

$$h_\ell(\beta_1, \ldots, \beta_{m+2}) = -\sum_{k=1}^{m} \tilde{w}_k \tilde{y}_k B_\ell(z_k) + \sum_{k=1}^{m} \tilde{w}_k \gamma(z_k) B_\ell(z_k) + \lambda \sum_{j=1}^{m+2} \beta_j \Omega_{j\ell}; \quad \ell = 1, \ldots, m+2.$$

To solve the equations $h_\ell(\beta_1, \ldots, \beta_{m+2}) = 0$; $\ell = 1, \ldots, m+2$; the Newton-Raphson method proceeds by iteratively solving the linear equation system, for the unknowns $\beta_1^{(n+1)}, \ldots, \beta_{m+2}^{(n+1)}$,

$$h_\ell\left(\beta_1^{(n)}, \ldots, \beta_{m+2}^{(n)}\right) + \sum_{j=1}^{m+2} \left(\beta_j^{(n+1)} - \beta_j^{(n)}\right) \frac{\partial h_\ell}{\partial \beta_j} \left(\beta_1^{(n)}, \ldots, \beta_{m+2}^{(n)}\right) = 0;$$
$$\ell = 1, \ldots, m+2.$$

We have

$$\frac{\partial h_\ell}{\partial \beta_j} = \sum_{k=1}^{m} \tilde{w}_k \gamma(z_k) B_j(z_k) B_\ell(z_k) + \lambda \Omega_{j\ell}.$$

Writing

$$\gamma_k^{(n)} = \exp\left\{\sum_{j=1}^{m+2} \beta_j^{(n)} B_j(z_k)\right\},$$

the set of linear equation becomes

$$-\sum_{k=1}^{m} \tilde{w}_k \tilde{y}_k B_\ell(z_k) + \sum_{k=1}^{m} \tilde{w}_k \gamma_k^{(n)} B_\ell(z_k) + \lambda \sum_{j=1}^{m+2} \beta_j^{(n)} \Omega_{j\ell}$$

$$+ \sum_{j=1}^{m+2} \left(\beta_j^{(n+1)} - \beta_j^{(n)}\right) \left(\sum_{k=1}^{m} \tilde{w}_k \gamma_k^{(n)} B_j(z_k) B_\ell(z_k) + \lambda \Omega_{j\ell}\right) = 0;$$

$$\ell = 1, \ldots, m+2.$$

Rewriting this, we get

$$\sum_{j=1}^{m+2} \sum_{k=1}^{m} \tilde{w}_k \gamma_k^{(n)} B_j(z_k) B_\ell(z_k) \beta_j^{(n+1)} + \lambda \sum_{j=1}^{m+2} \beta_j^{(n+1)} \Omega_{j\ell}$$

$$= \sum_{k=1}^{m} \tilde{w}_k \gamma_k^{(n)} \left(\left(\tilde{y}_k / \gamma_k^{(n)} - 1\right) + \sum_{j=1}^{m+2} \beta_j^{(n)} B_j(z_k)\right) B_\ell(z_k);$$

$$\ell = 1, \ldots, m+2. \tag{5.13}$$

Let $\mathbf{W}^{(n)}$ denote the diagonal matrix with elements $\tilde{w}_k \gamma_k^{(n)}$. Furthermore, let $\boldsymbol{\beta}^{(n)}$ denote the vector with elements $\beta_j^{(n)}$ and $\mathbf{y}^{(n)}$ the vector with elements $\tilde{y}_k / \gamma_k^{(n)} - 1 + \sum_{j=1}^{m+2} \beta_j^{(n)} B_j(z_k)$. The system of linear equations may then be written on matrix form as

$$(\mathbf{B}'\mathbf{W}^{(n)}\mathbf{B} + \lambda\boldsymbol{\Omega})\boldsymbol{\beta}^{(n+1)} = \mathbf{B}'\mathbf{W}^{(n)}\mathbf{y}^{(n)}. \tag{5.14}$$

We note that these are the same equations as we would get in the normal case with weights $\tilde{w}_k \gamma_k^{(n)}$ and observations $\tilde{y}_k / \gamma_k^{(n)} - 1 + \sum_{j=1}^{m+2} \beta_j^{(n)} B_j(z_k)$.

Figure 5.5 shows the same observed marginal claim frequencies for the age of the policyholder as in Fig. 5.1. The figure also shows a cubic spline fitted to the data by using the method just described for the Poisson case. The smoothing parameter λ was chosen by trial and error so as to obtain a good fit in the graph. However, one would of course prefer if λ could somehow be chosen automatically and we shall consider a method for doing this in Sect. 5.5.

5.3.3 Gamma Case

Finally, we turn to our main model for claim severity, the multiplicative gamma. As the calculations are very similar to the Poisson case, we omit some of the details.

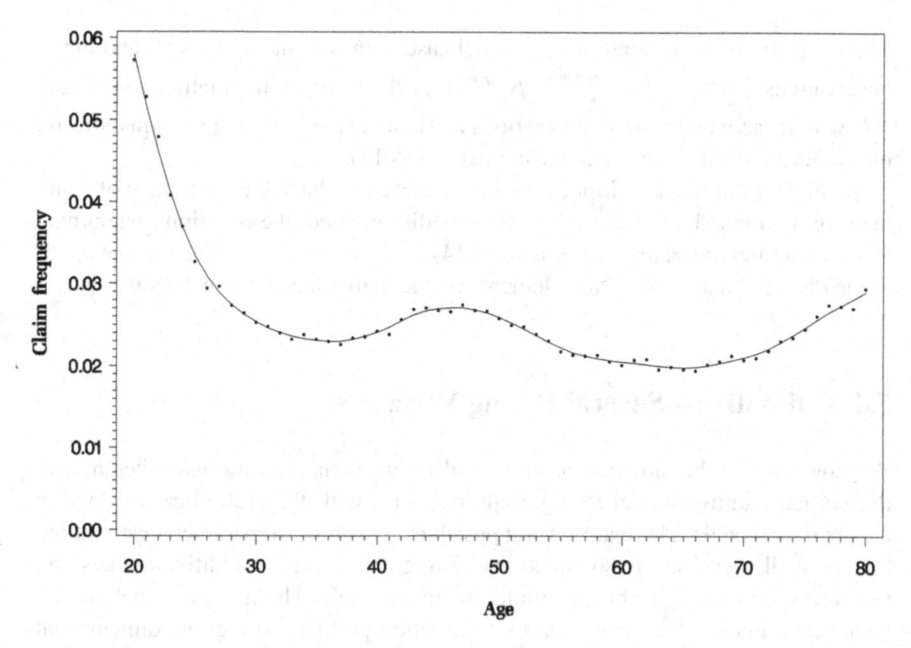

Fig. 5.5 Observed Motor TPL claim frequencies with fitted cubic spline

The link function is again $g(\mu) = \log \mu$ and the deviance is, cf. (3.4)

$$2\sum_i w_i(y_i/\mu_i - 1 - \log(y_i/\mu_i)). \tag{5.15}$$

Following the arguments and notation of the Poisson case, we arrive at the following equations, corresponding to (5.12):

$$-\sum_{k=1}^{m} \tilde{w}_k \frac{\tilde{y}_k}{\gamma(z_k)} B_\ell(z_k) + \sum_{k=1}^{m} \tilde{w}_k B_\ell(z_k) + \lambda \sum_{j=1}^{m+2} \beta_j \Omega_{j\ell} = 0;$$
$$\ell = 1, \ldots, m+2.$$

Again, these are nonlinear equations and applying the Newton-Raphson method, the equations corresponding to (5.13) now become

$$\sum_{j=1}^{m+2}\sum_{k=1}^{m} \tilde{w}_k \frac{\tilde{y}_k}{\gamma_k^{(n)}} B_j(z_k) B_\ell(z_k) \beta_j^{(n+1)} + \lambda \sum_{j=1}^{m+2} \beta_j^{(n+1)} \Omega_{j\ell}$$
$$= \sum_{k=1}^{m} \tilde{w}_k \frac{\tilde{y}_k}{\gamma_k^{(n)}} \left(1 - \frac{\gamma_k^{(n)}}{\tilde{y}_k} + \sum_{j=1}^{m+2} \beta_j^{(n)} B_j(z_k) \right) B_\ell(z_k);$$
$$\ell = 1, \ldots, m+2. \tag{5.16}$$

These equations correspond to the normal case with weights $\tilde{w}_k \tilde{y}_k / \gamma_k^{(n)}$ and the m observations $1 - \gamma_k^{(n)}/\tilde{y}_k + \sum_{j=1}^{m+2} \beta_j^{(n)} B_j(z_k)$. Introducing the matrices $\mathbf{W}^{(n)}$ and $\mathbf{y}^{(n)}$ with these weights and observations as elements, and writing the equations on matrix form, we again arrive at the expression (5.14).

As in the traditional multiplicative gamma case, we have the alternative of using the scoring method, cf. Sect. 3.2.3. As is readily verified, the equation system may again be written on matrix form as in (5.14). The vector $\mathbf{y}^{(n)}$ will be the same, but in the diagonal matrix $\mathbf{W}^{(n)}$ the elements on the main diagonal now become $\tilde{w}_k$.

5.4 Estimation—Several Rating Variables

We now turn to the situation with several rating variables and again begin with the normal additive model to highlight analogies with the multiplicative Poisson and gamma models. There are several possible algorithms for parameter estimation, but we shall focus on the so called backfitting algorithm for additive models and the corresponding algorithm for multiplicative models. The main idea behind the backfitting algorithm is to reduce the estimation problem to the one-dimensional case, discussed in the previous section. This idea also naturally leads to the concept of *partial residuals*, enabling the useful partial residual plots.

An alternative algorithm is the *penalized iteratively re-weighted least squares* scheme, see Wood [Wo06]. This is an extension of the algorithm in Sect. 3.2.3 and is usually faster than the simpler backfitting method.

5.4.1 Normal Case

Let us assume that we have a set of categorical variables and two continuous variables. The generalization to the case of an arbitrary number of continuous variables is completely straightforward.

We denote by x_{1i} the value of the first continuous variable in observation i and x_{2i} the value of the second continuous variable. Furthermore we use $z_{11}, \ldots, z_{1m_1}$ and $z_{21}, \ldots, z_{2m_2}$ to denote the set of possible values for these variables. We treat the categorical variables as in (5.3), and so the model is specified by

$$\eta_i = \sum_{j=0}^{r} \beta_j x'_{ij} + \sum_{k=1}^{m_1+2} \beta_{1k} B_{1k}(x_{1i}) + \sum_{\ell=1}^{m_2+2} \beta_{2\ell} B_{2\ell}(x_{2i}),$$

where $x'_{i0} = 1$ for all i, so that β_0 is the intercept. As always, the parameters are not uniquely determined without further restrictions. If we add some constant a_1 to all β_{1k} and some a_2 to all $\beta_{2\ell}$ while subtracting $a_1 + a_2$ from β_0, then due to the property (B.22) of B-splines, η_i remains unchanged. If we want to, we can achieve uniqueness by choosing base levels for the continuous variables. In practice, the rounding of the variable, for example rounding age to whole years or car weight to the nearest 50 kilograms, gives rise to a relatively small number of possible values. Therefore, we can choose some z_{1J} and z_{2K} as the base levels, where $J \in \{1, \ldots, m_1\}$ and $K \in \{1, \ldots, m_2\}$. This means that we have

$$\sum_{k=1}^{m_1+2} \beta_{1k} B_{1k}(z_{1J}) = 0 \quad \text{and} \quad \sum_{\ell=1}^{m_2+2} \beta_{2\ell} B_{2\ell}(z_{2K}) = 0. \tag{5.17}$$

We could now change the parametrization and eliminate one of the β_k as we have done for the categorical variables; however, we shall not pursue this possibility. It turns out that with the estimation procedure soon to be described, it is not necessary to put any restrictions on the β parameters. Arbitrary adjustments of the mean level can be made after the estimation is completed. In fact, in a tariff analysis, only the relativities are important; the mean level of the premium is decided based upon other considerations, cf. Sect. 3.6.7.

Remark 5.2 In cases where the number of values of a continuous variable occurring in the data is very large, one may want to reduce it to speed up execution. Either one can round the values to the nearest integer or whatever is appropriate, or one can use an interval subdivision and compute weighted averages of the values in the intervals to use as knots. In contrast with the traditional interval subdivision method, the choice of the number of intervals is not very important here. For an illustration of this phenomenon, see Grgić [Gr08, pp. 42–43].

In the additive normal case we have $\eta_i = \mu_i$ and we define the penalized deviance as

$$\Delta = \sum_i w_i (y_i - \mu_i)^2 + \lambda_1 \sum_{j=1}^{m_1+2} \sum_{k=1}^{m_1+2} \beta_{1j} \beta_{1k} \Omega_{jk}^{(1)} + \lambda_2 \sum_{j=1}^{m_2+2} \sum_{k=1}^{m_2+2} \beta_{2j} \beta_{2k} \Omega_{jk}^{(2)}. \tag{5.18}$$

The notation should be obvious. Thus we add two penalty terms, corresponding to the two continuous variables. Next, let us introduce the notation

$$\nu_{0i} = \sum_{j=0}^{r} \beta_j x'_{ij},$$

$$\nu_{1i} = \sum_{j=1}^{m_1+2} \beta_{1j} B_{1j}(x_{1i}),$$

$$\nu_{2i} = \sum_{j=1}^{m_2+2} \beta_{2j} B_{2j}(x_{2i}).$$

The deviance may then be written as

$$\begin{aligned}\sum_i w_i(y_i - \mu_i)^2 &= \sum_i w_i(y_i - (\nu_{0i} + \nu_{1i} + \nu_{2i}))^2 \\ &= \sum_i w_i((y_i - \nu_{1i} - \nu_{2i}) - \nu_{0i})^2. \end{aligned} \tag{5.19}$$

If $\beta_{11}, \ldots, \beta_{1,m_1+2}$ and $\beta_{21}, \ldots, \beta_{2,m_2+2}$ were known, the problem of estimating $\beta_0, \ldots, \beta_r$ would be identical to the usual problem with only categorical variables, since the penalty terms are then only constants, but with observations $y_i - \nu_{1i} - \nu_{2i}$. Similarly, if we knew $\beta_0, \ldots, \beta_r$ and $\beta_{21}, \ldots, \beta_{2,m_2+2}$, the problem of estimating $\beta_{11}, \ldots, \beta_{1,m_1+2}$ is precisely the one treated in Sect. 5.3.1, with the observations $y_i - \nu_{0i} - \nu_{2i}$. The corresponding statement of course holds for the second continuous variable.

These observations lie behind the following estimation method. First compute some initial estimates. For instance, $\hat{\beta}_0, \ldots, \hat{\beta}_r$ can be derived by analyzing the data with the continuous variables excluded and setting all $\{\hat{\beta}_{1j}\}$ and $\{\hat{\beta}_{2j}\}$ to zero. Given a set of estimates, put

$$\begin{aligned} \hat{\nu}_{0i} &= \sum_{j=0}^{r} \hat{\beta}_j x'_{ij}, \\ \hat{\nu}_{1i} &= \sum_{j=1}^{m_1+2} \hat{\beta}_{1j} B_{1j}(x_{1i}), \\ \hat{\nu}_{2i} &= \sum_{j=1}^{m_2+2} \hat{\beta}_{2j} B_{2j}(x_{2i}). \end{aligned} \tag{5.20}$$

Now proceed by iterating between the following three steps:

Step 1. Compute $\hat{\beta}_{11}, \ldots, \hat{\beta}_{1,m_1+2}$ as in Sect. 5.3.1 based on the "observations" $y_i - \hat{\nu}_{0i} - \hat{\nu}_{2i}$ with x_{1i} as a single explanatory variable.
Step 2. Compute $\hat{\beta}_{21}, \ldots, \hat{\beta}_{2,m_2+2}$ based on the "observations" $y_i - \hat{\nu}_{0i} - \hat{\nu}_{1i}$ with x_{2i} as a single explanatory variable.
Step 3. Compute new $\hat{\beta}_0, \ldots, \hat{\beta}_r$ based on the "observations" $y_i - \hat{\nu}_{1i} - \hat{\nu}_{2i}$ and using only the categorical variables.

One continues until the estimates have converged. This procedure is called the *backfitting algorithm* in Hastie and Tibshirani [HT90], which in addition contains results on the convergence of the algorithm, see also Green and Silverman [GS94].

Considering one of the continuous variables, for instance the first one, to update the estimates $\hat{\beta}_{11}, \ldots, \hat{\beta}_{1,m_1+2}$ we use the 'observations' $y'_i = y_i - \hat{\nu}_{0i} - \hat{\nu}_{2i}$. These are called the *partial residuals* corresponding to the first continuous variable. As in

Sect. 5.3.1, let I_{1k} denote the set of i for which $x_{1i} = z_{1k}$ and write

$$\tilde{w}_{1k} = \sum_{i \in I_{1k}} w_i, \qquad \tilde{y}_{1k} = \frac{1}{\tilde{w}_{1k}} \sum_{i \in I_{1k}} w_i y_i.$$

Now define $\tilde{y}'_{1k}$ by

$$\tilde{y}'_{1k} = \frac{1}{\tilde{w}_{1k}} \sum_{i \in I_{1k}} w_i y'_i = \frac{1}{\tilde{w}_{1k}} \sum_{i \in I_{1k}} w_i (y_i - \hat{\nu}_{0i} - \hat{\nu}_{2i}) = \tilde{y}_{1k} - \frac{1}{\tilde{w}_{1k}} \sum_{i \in I_{1k}} w_i \left(\hat{\nu}_{0i} + \hat{\nu}_{2i}\right).$$

We see that the equation corresponding to (5.8) now becomes

$$\sum_{k=1}^{m_1} \sum_{j=1}^{m_1+2} \tilde{w}_{1k} \beta_{1j} B_{1j}(z_{1k}) B_{1\ell}(z_{1k}) + \lambda_1 \sum_{j=1}^{m_1+2} \beta_{1j} \Omega^{(1)}_{j\ell}$$
$$= \sum_{k=1}^{m_1} \tilde{w}_{1k} \tilde{y}'_{1k} B_{1\ell}(z_{1k}); \quad \ell = 1, \ldots, m_1 + 2. \tag{5.21}$$

This is of course the same equation as we would have if we had a single continuous variable, weights $\tilde{w}_{11}, \ldots, \tilde{w}_{1m_1}$ and observations $\tilde{y}'_{11}, \ldots, \tilde{y}'_{1m_1}$. Thus, after convergence of the algorithm, we could check the fitted spline against these *aggregated partial residuals*. An example of such a partial residual plot will be given in the next subsection.

The backfitting algorithm and the possibility to define partial residuals are based on the fact that if we add a constant to a variable having a normal distribution, it remains normal. The scale invariance property (cf. Sect. 2.1.4) makes it possible to extend these concepts to multiplicative Tweedie models. We shall now verify this, beginning with the (relative) Poisson case.

5.4.2 *Poisson Case*

We now turn to the Poisson case and consider the same situation as in the previous section. However, instead of the ν_{ki} we now define

$$\gamma_{0i} = \exp\left\{\sum_{j=0}^{r} \beta_j x'_{ij}\right\},$$
$$\gamma_{1i} = \exp\left\{\sum_{j=1}^{m_1+2} \beta_{1j} B_{1j}(x_{1i})\right\}, \tag{5.22}$$
$$\gamma_{2i} = \exp\left\{\sum_{j=1}^{m_2+2} \beta_{2j} B_{2j}(x_{2i})\right\}.$$

Assume for the moment that the parameters $\beta_0, \ldots, \beta_r$ and $\beta_{21}, \ldots, \beta_{2,m_2+2}$ are known. Similarly to (5.19), the deviance in the Poisson case may be written as

$$
\begin{aligned}
&2\sum_i w_i(y_i \log y_i - y_i \log \mu_i - y_i + \mu_i) \\
&\quad = 2\sum_i w_i(y_i \log y_i - y_i \log(\gamma_{0i}\gamma_{1i}\gamma_{2i}) - y_i + \gamma_{0i}\gamma_{1i}\gamma_{2i}) \\
&\quad = 2\sum_i w_i \gamma_{0i}\gamma_{2i}\left(\frac{y_i}{\gamma_{0i}\gamma_{2i}} \log \frac{y_i}{\gamma_{0i}\gamma_{2i}} - \frac{y_i}{\gamma_{0i}\gamma_{2i}} \log \gamma_{1i} - \frac{y_i}{\gamma_{0i}\gamma_{2i}} + \gamma_{1i}\right).
\end{aligned}
$$

We may then define

$$
w_i' = w_i \gamma_{0i}\gamma_{2i}, \qquad y_i' = \frac{y_i}{\gamma_{0i}\gamma_{2i}},
$$

and the penalized deviance becomes

$$
\Delta = 2\sum_i w_i'(y_i' \log y_i' - y_i' \log \gamma_{1i} - y_i' + \gamma_{1i}) + \lambda_1 \sum_{j=1}^{m_1+2} \sum_{k=1}^{m_1+2} \beta_{1j}\beta_{1k}\Omega_{jk}^{(1)}, \quad (5.23)
$$

where we have excluded the second penalty term since it is just a constant when estimating $\beta_{11}, \ldots, \beta_{1,m_1+2}$. Since the deviance is symmetric in γ_{0i}, γ_{1i} and γ_{2i}, we can factor out $\gamma_{0i}\gamma_{1i}$ or $\gamma_{1i}\gamma_{2i}$ in the same way.

We now realize that we may define a backfitting algorithm in analogy with the normal case. Similarly to (5.20), we define

$$
\begin{aligned}
\hat{\gamma}_{0i} &= \exp\left\{\sum_{j=0}^{r} \hat{\beta}_j x_{ij}'\right\}, \\
\hat{\gamma}_{1i} &= \exp\left\{\sum_{j=1}^{m_1+2} \hat{\beta}_{1j} B_{1j}(x_{1i})\right\}, \qquad (5.24) \\
\hat{\gamma}_{2i} &= \exp\left\{\sum_{j=1}^{m_2+2} \hat{\beta}_{2j} B_{2j}(x_{2i})\right\}.
\end{aligned}
$$

The backfitting algorithm in this case consists of iterating between the following three steps:

Step 1. Compute $\hat{\beta}_{11}, \ldots, \hat{\beta}_{1,m_1+2}$ as in Sect. 5.3.2 based on the "observations" $y_i/(\hat{\gamma}_{0i}\hat{\gamma}_{2i})$ and weights $w_i\hat{\gamma}_{0i}\hat{\gamma}_{2i}$, with x_{1i} as a single explanatory variable.
Step 2. Compute $\hat{\beta}_{21}, \ldots, \hat{\beta}_{2,m_2+2}$ based on the "observations" $y_i/(\hat{\gamma}_{0i}\hat{\gamma}_{1i})$ and weights $w_i\hat{\gamma}_{0i}\hat{\gamma}_{1i}$, with x_{2i} as a single explanatory variable.
Step 3. Compute new $\hat{\beta}_0, \ldots, \hat{\beta}_r$ based on the "observations" $y_i/(\hat{\gamma}_{1i}\hat{\gamma}_{2i})$, weights $w_i\hat{\gamma}_{1i}\hat{\gamma}_{2i}$, and using only the categorical variables.

Each step amounts to using a GLM with an offset, cf. Sect. 3.6.3. In fact, this procedure is entirely similar to the backfitting algorithm in Sect. 4.2.2, and this is why we chose the same name there. With I_{1k} as in the previous subsection, we define

$$\tilde{w}'_{1k} = \sum_{i \in I_{1k}} w'_i, \qquad \tilde{y}'_{1k} = \frac{1}{\tilde{w}'_{1k}} \sum_{i \in I_{1k}} w'_i y'_i.$$

Proceeding as in the normal case, (5.23) may be written

$$2\sum_{k=1}^{m_1} \tilde{w}'_{1k}(\tilde{y}'_{1k} \log \tilde{y}'_{1k} - \tilde{y}'_{1k} \log \mu_1(z_{1k}) - \tilde{y}'_{1k} + \mu_1(z_{1k}))$$
$$+ \lambda_1 \sum_{j=1}^{m_1+2} \sum_{k=1}^{m_1+2} \beta_{1j}\beta_{1k}\Omega^{(1)}_{jk},$$

where

$$\mu_1(x) = \exp\left\{\sum_{k=1}^{m_1+2} \beta_{1k} B_{1k}(x)\right\}.$$

Thus what we are doing is fitting $\mu_1(z_{11}), \ldots, \mu_1(z_{1m_1})$ to the "observations" $\tilde{y}'_{11}, \ldots, \tilde{y}'_{1,m_1}$. A natural graphical method for checking the fit after completing the estimation of the parameters would therefore be to plot $\mu_1(x)$ with inserted $\hat{\beta}_{11}, \ldots, \hat{\beta}_{1m_1}$ against $\tilde{y}'_{11}, \ldots, \tilde{y}'_{1m_1}$. The values $\tilde{y}'_{11}, \ldots, \tilde{y}'_{1m_1}$ may be called *aggregated partial residuals* in analogy with the normal case.

An example of a partial residual plot, borrowed from Grgić [Gr08], is shown in Fig. 5.6. It concerns our basic moped data from Example 1.1, with the difference that the variable vehicle age is treated as a continuous variable instead of being categorized into just two age classes. The figure shows the fitted spline for the claim frequency against the aggregated partial residuals. A partial residual plot can be used to validate the fit of the spline to the data, and in particular to choose the value of the smoothing parameter λ. Here though, λ was selected using cross validation, as explained in Sect. 5.5.

The methodology described so far covers the case of an arbitrary number of categorical and/or continuous rating factors. In case we also have a multi-class factor, the steps of the backfitting algorithm described above can be inserted in place of Step 1 in the backfitting algorithm described in Sect. 4.2.2. The details are left to the interested reader.

Remark 5.3 As already mentioned, one may introduce a base level for a continuous rating factor. For instance, in the example of Fig. 5.6, two year old mopeds were chosen as the base level. Assuming vehicle age to be the first continuous variable in

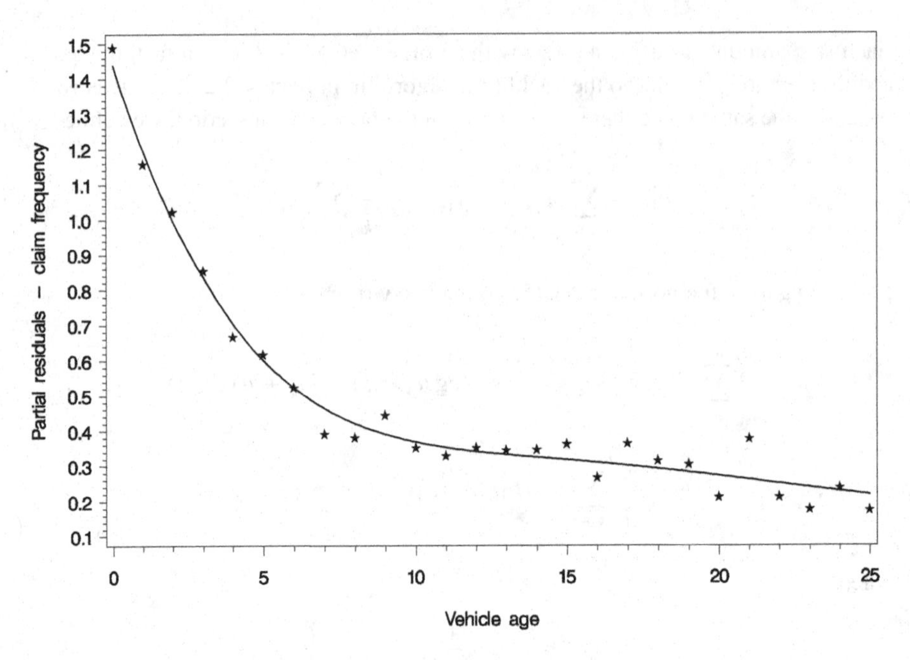

Fig. 5.6 Partial residual plot from moped data

the above discussion, let us write

$$g_{1k} = \exp\left\{\sum_{j=1}^{m_1+2} \hat{\beta}_{1j} B_{1j}(z_k)\right\}, \qquad g_0 = \exp\{\hat{\beta}_0\}.$$

If z_J is to be the base level, we divide all g_{1k} by g_{1J} and multiply g_0 by g_{1J}.

Remark 5.4 Sometimes it is required that a certain interval subdivision be maintained for a continuous variable. For instance, it may be complicated to change the structure of the tariff tables and one has to stick to three year intervals for vehicle age. Still, one may carry out the analysis using an arbitrary set of knots and then use the fitted spline to compute relativities for the intervals, by suitably weighing together the relativities for the knots.

5.4.3 Gamma Case

We continue with the same notation as in the previous subsection. The deviance in the gamma case is, cf. Sect. 3.1,

$$D = 2\sum_i w_i(y_i/\mu_i - 1 - \log(y_i/\mu_i)). \tag{5.25}$$

Using the same notation as for the Poisson case in (5.22) and (5.4.2), the deviance may be written

$$2\sum_i w_i\left(\frac{y_i}{\gamma_{0i}\gamma_{1i}\gamma_{2i}} - 1 - \log\frac{y_i}{\gamma_{0i}\gamma_{1i}\gamma_{2i}}\right)$$

$$= 2\sum_i w_i\left(\frac{y_i'}{\gamma_{2i}} - 1 - \log\frac{y_i'}{\gamma_{2i}}\right).$$

We see again that if $\beta_0, \ldots, \beta_r$ and $\beta_{1i}, \ldots, \beta_{1m_1}$ were known, so that γ_{0i} and γ_{1i} were also known, the deviance is the same as in the one variable case when estimating $\beta_{21}, \ldots, \beta_{2,m_2}$. In the gamma case, as in the normal case, the weights need not be modified. Using this, the algorithm in the gamma case can be constructed following the steps in the Poisson case.

5.5 Choosing the Smoothing Parameter

Depending on the smoothing parameter λ, the fitted cubic spline varies between two extremes: a straight line and the interpolating natural cubic spline with perfect fit to the data. The question remains: how do we determine λ? As mentioned in Sect. 5.4.2, a simple way of choosing λ is to plot the fitted cubic spline for various values of λ and compare with the partial residuals. However, this is a subjective procedure and there is also a need for a data-based automatic selection of λ. In this section we describe the method known as *cross-validation*.

We begin with the case of one rating variable and the normal distribution. The fitted cubic spline minimizes the expression

$$\Delta(s) = \sum_{k=1}^{m} \tilde{w}_k(\tilde{y}_k - s(z_k))^2 + \lambda\int_a^b (s''(x))^2\,dx, \tag{5.26}$$

where the smoothing parameter λ controls the variability s. The idea behind cross-validation is to choose λ as follows.

Suppose we delete a certain z_k and the corresponding $\tilde{y}_k$ from the data. We can then, for any λ, compute the minimizing cubic spline $s_k^\lambda(x)$ for this smaller data set. Now, with an appropriate value of λ, the estimated mean at z_k, i.e. $s_k^\lambda(z_k)$, should be a good predictor of the deleted data point $\tilde{y}_k$. Furthermore, this should be the case for any k. It is natural to measure this overall ability to predict the deleted observations by the corresponding deviance

$$C(\lambda) = \sum_{k=1}^{m} \tilde{w}_k\left(\tilde{y}_k - s_k^\lambda(z_k)\right)^2, \tag{5.27}$$

in the same way as we have previously measured the fit of a set of estimated means to the data. The idea of cross-validation is to choose λ as the value for which the *cross validation score* $C(\lambda)$ is minimized.

It appears that to compute $C(\lambda)$, we would have to solve m minimization problems, one for each deleted observation. However, we shall now show that in the normal case, the computation of $C(\lambda)$ can be simplified.

First, recall that for a given λ and any observation vector $\mathbf{y}$ with m elements, the coefficients $\beta_1, \ldots, \beta_{m+2}$ are given as the solution to the linear equation system (5.9):

$$\left(\mathbf{B}'\mathbf{W}\mathbf{B} + \lambda\boldsymbol{\Omega}\right)\beta = \mathbf{B}'\mathbf{W}\mathbf{y}.$$

Let $\hat{\mu}_k$ denote the estimated mean at z_k, i.e. $\hat{\mu}_k = \sum_{j=1}^{m+2} \hat{\beta}_j B_j(z_k)$. If $\hat{\boldsymbol{\mu}}$ denotes the column vector with elements $\{\hat{\mu}_k\}$, we have $\hat{\boldsymbol{\mu}} = \mathbf{B}\hat{\boldsymbol{\beta}}$. It follows that whatever the observations in the vector $\mathbf{y}$ are, the estimated means are given by $\hat{\boldsymbol{\mu}} = \mathbf{A}\mathbf{y}$, where

$$\mathbf{A} = \mathbf{B}(\mathbf{B}'\mathbf{W}\mathbf{B} + \lambda\boldsymbol{\Omega})^{-1}\mathbf{B}'\mathbf{W}. \tag{5.28}$$

Now, let λ be given and let $s_k(x) = s_k^\lambda(x)$ be the minimizing cubic spline for the case when we have deleted the kth observation, where we omit the superindex λ for notational convenience below. Consider the vector $\mathbf{y}^*$ of length m, with components y_j^* given by $y_j^* = \tilde{y}_j$, if $j \neq k$, and $y_k^* = s_k(z_k)$. Then for any twice differentiable function $f(x)$,

$$\begin{aligned}
&\sum_{j=1}^{m} \tilde{w}_j \left(y_j^* - f(z_j)\right)^2 + \lambda \int_a^b (f''(x))^2\,dx \\
&\geq \sum_{\substack{1\leq j\leq m\\ j\neq k}} \tilde{w}_j \left(y_j^* - f(z_j)\right)^2 + \lambda \int_a^b (f''(x))^2\,dx \\
&\geq \sum_{\substack{1\leq j\leq m\\ j\neq k}} \tilde{w}_j \left(y_j^* - s_k(z_j)\right)^2 + \lambda \int_a^b (s_k''(x))^2\,dx \\
&= \sum_{j=1}^{m} \tilde{w}_j \left(y_j^* - s_k(z_j)\right)^2 + \lambda \int_a^b (s_k''(x))^2\,dx.
\end{aligned}$$

This shows that for the observations $y_1^*, \ldots, y_m^*$, the minimizing cubic spline is $s_k(x)$. Thus if $s(x)$ denotes the minimizing cubic spline for the observations $\tilde{y}_1, \ldots, \tilde{y}_m$, and $\mathbf{A} = \{a_{kj}\}$, we have

$$\begin{aligned}
s_k(z_k) &= \sum_{j=1}^{m} a_{kj} y_j^* = \sum_{\substack{1\leq j\leq m\\ j\neq k}} a_{kj}\tilde{y}_j + a_{kk} y_k^* \\
&= \sum_{j=1}^{m} a_{kj}\tilde{y}_j - a_{kk}\tilde{y}_k + a_{kk} s_k(z_k) \\
&= s(z_k) - a_{kk}\tilde{y}_k + a_{kk} s_k(z_k).
\end{aligned}$$

Thus,

$$\tilde{y}_k - s_k(z_k) = \tilde{y}_k - s(z_k) + a_{kk}(\tilde{y}_k - s_k(z_k)), \tag{5.29}$$

and so

$$\tilde{y}_k - s_k(z_k) = \frac{\tilde{y}_k - s(z_k)}{1 - a_{kk}}. \tag{5.30}$$

We have now shown that

$$C(\lambda) = \sum_{k=1}^{m} \tilde{w}_k \left(\frac{\tilde{y}_k - s(z_k)}{1 - a_{kk}} \right)^2, \tag{5.31}$$

and so $C(\lambda)$ can be computed from the estimated means $s(z_1), \ldots, s(z_m)$ for the full data set, together with the diagonal elements of the matrix $\mathbf{A}$. A minimizing λ can then be found by some appropriate search method.

In analogy with the normal case, we would define the cross-validation score in the Poisson case using the deviance

$$C(\lambda) = 2 \sum_{k=1}^{m} \tilde{w}_k (\tilde{y}_k \log \tilde{y}_k - \tilde{y}_k s_k^{\lambda}(z_k) - \tilde{y}_k + \exp\{s_k^{\lambda}(z_k)\}). \tag{5.32}$$

However, looking through the derivation of the simplification (5.31), we see that it relies heavily on the linearity of the normal case, and so it does not generalize to the Poisson case. Of course, we may compute $C(\lambda)$ by finding all the minimizing splines $s_1^{\lambda}, \ldots, s_m^{\lambda}$, but this could be time consuming. Because of this, an approximative method has been suggested, which goes as follows.

Recall that in each iteration of the Newton-Raphson method, we are solving the system of linear equations (5.14), which may be viewed as having arisen from normal observations $\tilde{y}_k / \gamma_k^{(n)} - 1 + \sum_{j=1}^{m+2} \beta_j^{(n)} B_j(z_k)$, and with weights $\tilde{w}_k \gamma_k^{(n)}$. The idea is to obtain the minimizing spline $s^{(n)}(x)$ for this situation and the corresponding matrix

$$\mathbf{A}^{(n)} = \mathbf{B}(\mathbf{B}'\mathbf{W}^{(n)}\mathbf{B} + \lambda\boldsymbol{\Omega})^{-1}\mathbf{B}'\mathbf{W}^{(n)} \tag{5.33}$$

and obtain the approximate cross validation score

$$C^{(n)}(\lambda) = \sum_{k=1}^{m} \tilde{w}_k \left(\frac{\tilde{y}_k - s^{(n)}(x)}{1 - a_{kk}^{(n)}} \right)^2. \tag{5.34}$$

The minimizing λ is then used in the next iteration, until convergence.

In the case of several rating variables, a simple approach is to proceed as in Sect. 5.4.2, and apply the above method each time we update the estimates for a particular continuous variable. Usually, this method gives reasonable estimates of the smoothing parameter, which can be checked using the partial residuals. For more refined estimation methods, see Sects. 4.5–4.7 in Wood [Wo06].

The computation of the matrix $\mathbf{A}$ defined in (5.28) involves computing the inverse of the matrix $\mathbf{B}'\mathbf{W}\mathbf{B} + \lambda\boldsymbol{\Omega}$. However from (5.31) we see that only the diagonal elements of $\mathbf{A}$ are used in the cross-validation score. There is a fast algorithm for computing these diagonal elements without matrix inversion, making use of the banded structure of $\mathbf{B}$ and $\boldsymbol{\Omega}$. This algorithm is described in Sect. 3.2.2 of Green and Silverman [GS94], see also the errata for that text.

Depending on the scale for the continuous variable, the minimizing λ can sometimes be extremely large. It is usually a good idea to rescale these variable values together with the weights, see e.g. Grgić [Gr08, p. 24]. That paper also contains some interesting non-life insurance examples on cross-validation, for example situations where the function $C(\lambda)$ have several minima.

5.6 Interaction Between a Continuous and a Categorical Variable

How to model interaction between two categorical variables has been discussed in Sect. 3.6.2. In this section we treat the case when one of the variables is categorical and the other continuous. The case when both variable are continuous, which involves the concept of bivariate splines, is treated in the next section.

Let us first consider the case where we have only the two interacting variables and no other. We let x_{1i} denote the value of the categorical variable for the ith observation, and x_{2i} the value of the continuous variable. We denote by $z_{11}, \ldots, z_{1m_1}$ and $z_{21}, \ldots, z_{2m_2}$ the possible values of x_{1i} and x_{2i}, respectively. Define the functions $\phi_j(\cdot)$, $j = 1, \ldots, m_1$ by setting $\phi_j(x) = 1$, if $x = z_{1j}$, and $\phi_j(x) = 0$, otherwise. Furthermore, let $B_1(\cdot), \ldots, B_{m_2+2}(\cdot)$ denote the cubic B-splines for the knots $z_{21}, \ldots, z_{2m_2}$.

It is natural to model the effect by m_1 different splines, one for each possible value of x_{1i}. In the example of age effects for male and female drivers in Sect. 3.6.2 there would be one spline for men and one for women. Thus we may express our model as

$$\eta_i = \eta(x_{1i}, x_{2i}) = \sum_{j=1}^{m_1} \sum_{k=1}^{m_2+2} \beta_{jk}\phi_j(x_{1i})B_k(x_{2i}).$$

Let $s_j(x)$ denote the cubic spline $\sum_{k=1}^{m_2+2} \beta_{jk} B_k(x)$. Since $\phi_j(z_{1j}) = 1$, we see that if $x_{1i} = z_{1j}$, then $\eta_i = s_j(x_{2i})$.

Consider now the Poisson case and write $\mu(x_{1i}, x_{2i}) = \exp\{\eta(x_{1i}, x_{2i})\}$. In analogy with the corresponding expression for a single continuous rating variable, the penalized deviance is defined as

$$2\sum_i w_i(y_i \log y_i - y_i \log\mu(x_{1i}, x_{2i}) - y_i + \mu(x_{1i}, x_{2i})) + \sum_{j=1}^{m_1} \lambda_j \int (s_j''(x))^2\, dx.$$

Since the effect of the continuous variable may be quite different for different values of the categorical variable, as may also the data volume, it is natural to permit individual smoothing parameters $\lambda_1, \ldots, \lambda_{m_1}$.

Next, for $j = 1, \ldots, m_1$ and $k = 1, \ldots, m_2$, we denote by I_{jk} the set of i for which $x_{1i} = z_{1j}$ and $x_{2i} = z_{2k}$. We then define $\tilde{w}_{jk}$ and $\tilde{y}_{jk}$ by

$$\tilde{w}_{jk} = \sum_{i \in I_{jk}} w_i, \qquad \tilde{y}_{jk} = \frac{1}{\tilde{w}_{jk}} \sum_{i \in I_{jk}} w_i y_i.$$

Differentiating the penalized deviance with respect to β_{rs} and setting the derivatives equal to zero, we obtain for each $r = 1, \ldots, m_1$,

$$-\sum_{k=1}^{m_2} \tilde{w}_{rk} \tilde{y}_{rk} B_s(z_{2k}) + \sum_{k=1}^{m_2} \tilde{w}_{rk} \exp\left\{\sum_{\ell=1}^{m_2+2} \beta_{r\ell} B_\ell(z_{2k})\right\} B_s(z_{2k})$$

$$+ \lambda_r \sum_{\ell=1}^{m_2+2} \beta_{r\ell} \Omega_{\ell s} = 0; \quad s = 1, \ldots, m_2 + 2;$$

recall the definition of Ω_{jk} in (5.3.1). Comparing with (5.12), we see that our present problem is the same as if we had m_1 different sets of data, the r:th having weights $\tilde{w}_{rs}$ and observations $\tilde{y}_{rs}$, $s = 1, \ldots, m_2$. Thus we may carry out the fitting of the m_1 splines independently of one another. The gamma case is entirely similar.

Having additional explanatory variables involves nothing new and one may iterate between fitting the above splines and the fitting connected with the other variables, as described in Sect. 5.4.2.

For an application of the method described in this section, see Grgić [Gr08, p. 48].

5.7 Bivariate Splines

In Sect. 5.6 we studied how to handle interactions between a continuous variable and a categorical one. We now consider the situation where we have two interacting continuous variables. This involves the fitting of some smooth surface to observations corresponding to points in R^2. Generalizing the concept of penalized deviances, we shall see that it is possible to derive explicit expressions for the solutions to the resulting minimization problem. It turns out that these are not piecewise polynomials and that some of the nice properties of the one-dimensional case are lost. Still, we will have a useful method for modelling interactions between two continuous variables.

5.7.1 Thin Plate Splines

We now assume that we have two continuous variables and let (x_{1i}, x_{2i}) denote the values of the pair of variables for observation i. Furthermore, we let $(z_{11}, z_{21}), \ldots, (z_{1m}, z_{2m})$ denote all possible values of the pair of variables. Note that this is not necessarily the set of all possible combinations of the values assumed by the separate variables. For certain values of the first variable, not all values of the second variable may occur in the data.

We want to model the effect of the variables by means of a function $f(x_1, x_2)$, occurring in the deviance through

$$\eta_i = f(x_{1i}, x_{2i}) + \cdots.$$

Let us denote the corresponding deviance by $D(f)$ to highlight the dependency on f. Obviously, two functions f and g satisfying $f(z_{1k}, z_{2k}) = g(z_{1k}, z_{2k})$ for $k = 1, \ldots, m$ also satisfies $D(f) = D(g)$.

It is not immediately obvious what the penalty term should look like in two dimensions. However, a natural choice would be to put restraints on all of the second partial derivatives and use

$$\int\!\!\int \left(\left(\frac{\partial^2 f}{\partial x_1^2} \right)^2 + \left(\frac{\partial^2 f}{\partial x_1 \partial x_2} \right)^2 + \left(\frac{\partial^2 f}{\partial x_2 \partial x_1} \right)^2 + \left(\frac{\partial^2 f}{\partial x_2^2} \right)^2 \right) dx_1\, dx_2,$$

that is

$$\int\!\!\int \left(\left(\frac{\partial^2 f}{\partial x_1^2} \right)^2 + 2 \left(\frac{\partial^2 f}{\partial x_1 \partial x_2} \right)^2 + \left(\frac{\partial^2 f}{\partial x_2^2} \right)^2 \right) dx_1\, dx_2. \tag{5.35}$$

In the one-dimensional case the integral was defined over an arbitrary interval containing all the observation points, and we could also have integrated over R. In two dimensions, it will not be immaterial what integration region we choose. Integrating over finite regions is more complicated, and therefore we integrate over the entire R^2.

It turns out that we will also have to make some changes to the definition of the family $\mathcal{F}$ of permissible functions. We would like to take the set of twice continuously differentiable functions, but this turns out to be a bit too strong. Of course, the integral in (5.35) must be well-defined, but for things to work out we require only that the second partial derivatives are square integrable. Furthermore, it will also be practical to require that the first partial derivatives are bounded, which anyway seems like a nice property of the type of functions we are looking for. The reasons for why exactly these requirements are appropriate are made clear in Appendix B.3. So we define $\mathcal{F}$ to be the set of functions f on R^2, such that

(i) The first partial derivatives of f are continuous and bounded on R;
(ii) The second partial derivatives of f are continuous for all $(x_1, x_2) \neq (z_{1j}, z_{2j})$; $j = 1, \ldots, m$

(iii) The integral (5.35) is well-defined and finite.

In analogy with the univariate case, we want to find an $f \in \mathcal{F}$ minimizing the penalized deviance

$$\Delta(f) = D(f) + \lambda \iint \left(\left(\frac{\partial^2 f}{\partial x_1^2} \right)^2 + 2 \left(\frac{\partial^2 f}{\partial x_1 \partial x_2} \right)^2 + \left(\frac{\partial^2 f}{\partial x_2^2} \right)^2 \right) dx_1\, dx_2.$$

In Appendix B.3 it is shown that a solution is given by

$$s(x_1, x_2) = \frac{1}{16\pi} \sum_{k=1}^{m} d_k r_k^2(x_1, x_2) \log r_k^2(x_1, x_2) + a_0 + a_1 x_1 + a_2 x_2, \tag{5.36}$$

where $r_k(x_1, x_2) = \sqrt{(x_1 - z_{1k})^2 + (x_2 - z_{2k})^2}$ and the parameters $d_1, \ldots, d_m$ satisfy

$$\sum_{k=1}^{m} d_k = 0, \qquad \sum_{k=1}^{m} d_k z_{1k} = 0, \qquad \sum_{k=1}^{m} d_k z_{2k} = 0. \tag{5.37}$$

Furthermore, it is shown that the penalty term can be written as

$$\begin{aligned} &\iint \left(\left(\frac{\partial^2 f}{\partial x_1^2} \right)^2 + 2 \left(\frac{\partial^2 f}{\partial x_1 \partial x_2} \right)^2 + \left(\frac{\partial^2 f}{\partial x_2^2} \right)^2 \right) dx_1\, dx_2 \\ &= \frac{1}{16\pi} \sum_{j=1}^{m} \sum_{k=1}^{m} d_j d_k r_k^2(z_{1j}, z_{2j}) \log r_k^2(z_{1j}, z_{2j}). \end{aligned}$$

5.7.2 Estimation with Thin Plate Splines

We have used the B-spline representation of univariate splines, but there also exists a representation similar to (5.36). In Appendix B.1 it is shown that a natural cubic spline with knots $u_1, \ldots, u_m$ can be written as

$$s(x) = \frac{1}{12} \sum_{k=1}^{m} d_k |x - u_k|^3 + a_0 + a_1 x, \tag{5.38}$$

for some constants $d_1, \ldots, d_m, a_0, a_1$, where $d_1, \ldots, d_m$ satisfy the conditions

$$\sum_{k=1}^{m} d_k = 0, \qquad \sum_{k=1}^{m} d_k u_k = 0. \tag{5.39}$$

Before continuing with thin plate splines, let us see what form the estimation problem takes using the representation (5.38) of natural cubic splines. We begin with one continuous rating variable and assuming a normal distribution. The notation is the same as in Sect. 5.3.1.

In Appendix B.1 (Lemmas B.3 and B.4), it is shown that the penalized deviance (5.4) takes the form

$$\begin{aligned}\Delta &= \sum_i w_i(y_i - s(x_i))^2 + \lambda \int_{-\infty}^{\infty} (s''(x))^2\,dx \\ &= \sum_i w_i \left(y_i - \left(\frac{1}{12} \sum_{j=1}^{m} d_j |x_i - z_j|^3 + a_1 + a_2 x_i \right) \right)^2 \\ &\quad + \lambda \frac{1}{12} \sum_{j=1}^{m} \sum_{k=1}^{m} d_j |z_j - z_k|^3.\end{aligned}$$

As before in the normal case, we have a linear minimization problem, but under the constraints (5.39). If we put

$$\begin{aligned}\phi_1(d_1, \ldots, d_m, a_1, a_2) &= d_1 + \cdots + d_m, \\ \phi_2(d_1, \ldots, d_m, a_1, a_2) &= d_1 z_1 + \cdots + d_m z_m,\end{aligned}$$

the problem is to minimize Δ subject to the conditions

$$\begin{aligned}\phi_1(d_1, \ldots, d_m, a_1, a_2) &= 0, \\ \phi_2(d_1, \ldots, d_m, a_1, a_2) &= 0.\end{aligned}$$

A standard method for solving problems of this kind is the Lagrange multiplier method. Introducing Lagrange multipliers α_1 and α_2, the minimizing $d_1, \ldots, d_m$, a_1, a_2 may be found by solving the equations

$$\begin{aligned}&\frac{\partial \Delta}{\partial d_r} + \alpha_1 \frac{\partial \phi_1}{\partial d_r} + \alpha_2 \frac{\partial \phi_2}{\partial d_r} = 0; \quad r = 1, \ldots, m; \\ &\frac{\partial \Delta}{\partial a_1} + \alpha_1 \frac{\partial \phi_1}{\partial a_1} + \alpha_2 \frac{\partial \phi_2}{\partial a_1} = 0; \\ &\frac{\partial \Delta}{\partial a_2} + \alpha_1 \frac{\partial \phi}{\partial a_2} + \alpha_2 \frac{\partial \psi}{\partial a_2} = 0; \\ &\phi_1(d_1, \ldots, d_m, a_1, a_2) = 0; \\ &\phi_2(d_1, \ldots, d_m, a_1, a_2) = 0.\end{aligned} \tag{5.40}$$

Recalling the notation $\tilde{w}_k$ and $\tilde{y}_k$ from Sect. 5.3.1, and introducing $e_{jk} = |z_j - z_k|^3/12$, these equations become

$$\sum_{j=1}^{m}\sum_{k=1}^{m} \tilde{w}_k e_{kj} d_j e_{kr} + \lambda \sum_{k=1}^{m} d_k e_{kr} + a_1 \sum_{k=1}^{m} \tilde{w}_k e_{kr}$$
$$+ a_2 \sum_{k=1}^{m} \tilde{w}_k z_k e_{kr} + \alpha_1 + \alpha_2 z_r = \sum_{k=1}^{m} \tilde{w}_k \tilde{y}_k e_{kr}; \quad r = 1, \ldots, m;$$
$$\sum_{j=1}^{m}\sum_{k=1}^{m} \tilde{w}_k d_j e_{jk} + a_1 \sum_{k=1}^{m} \tilde{w}_k + a_2 \sum_{k=1}^{m} \tilde{w}_k z_k = \sum_{k=1}^{m} \tilde{w}_k \tilde{y}_k; \tag{5.41}$$
$$\sum_{j=1}^{m}\sum_{k=1}^{m} \tilde{w}_k d_j z_k e_{jk} + a_1 \sum_{k=1}^{m} \tilde{w}_k z_k + a_2 \sum_{k=1}^{m} \tilde{w}_k z_k^2 = \sum_{k=1}^{m} \tilde{w}_k z_k \tilde{y}_k;$$
$$\sum_{k=1}^{m} d_k = 0;$$
$$\textstyle\sum_{k=1}^{m} d_k z_k = 0.$$

We now want to write this on matrix form. It will be convenient to introduce the notation $\mathbf{x} * \mathbf{y}$ for the componentwise multiplication of two vectors $\mathbf{x}$ and $\mathbf{y}$:

$$\mathbf{x} = \begin{bmatrix} x_1 \\ x_2 \\ \vdots \\ x_m \end{bmatrix}, \qquad \mathbf{y} = \begin{bmatrix} y_1 \\ y_2 \\ \vdots \\ y_m \end{bmatrix}, \qquad \mathbf{x} * \mathbf{y} = \begin{bmatrix} x_1 y_1 \\ x_2 y_2 \\ \vdots \\ x_m y_m \end{bmatrix}.$$

We also use the same notation with a vector and a matrix:

$$\mathbf{A} = \begin{bmatrix} a_{11} & a_{12} & \cdots & a_{1n} \\ a_{21} & a_{22} & \cdots & a_{2n} \\ \vdots & \vdots & & \vdots \\ a_{m1} & a_{m2} & \cdots & a_{mn} \end{bmatrix}, \qquad \mathbf{x} * \mathbf{A} = \begin{bmatrix} x_1 a_{11} & x_1 a_{12} & \cdots & x_1 a_{1n} \\ x_2 a_{21} & x_2 a_{22} & \cdots & x_1 a_{2n} \\ \vdots & \vdots & & \vdots \\ x_m a_{m1} & x_m a_{m2} & \cdots & x_m a_{mn} \end{bmatrix}.$$

Introduce the vectors

$$\mathbf{w} = \begin{bmatrix} \tilde{w}_1 \\ \tilde{w}_2 \\ \vdots \\ \tilde{w}_m \end{bmatrix}, \qquad \mathbf{y} = \begin{bmatrix} \tilde{y}_1 \\ \tilde{y}_2 \\ \vdots \\ \tilde{y}_m \end{bmatrix}, \qquad \mathbf{z} = \begin{bmatrix} z_1 \\ z_2 \\ \vdots \\ z_m \end{bmatrix},$$

and let $\mathbf{E}$ be the matrix with elements e_{jk}. Using block matrices, and using "$\cdot$" to denote matrix multiplication, we may now write (5.41) as

$$\begin{bmatrix} \mathbf{E}\cdot(\mathbf{E}*\mathbf{w})+\lambda\mathbf{E} & \mathbf{E}\cdot\mathbf{w} & \mathbf{E}\cdot(\mathbf{w}*\mathbf{z}) & \mathbf{1} & \mathbf{z} \\ (\mathbf{E}\cdot\mathbf{w})' & \mathbf{w}'\cdot\mathbf{1} & \mathbf{w}'\cdot\mathbf{z} & 0 & 0 \\ (\mathbf{E}\cdot(\mathbf{w}*\mathbf{z}))' & \mathbf{w}'\cdot\mathbf{z} & \mathbf{w}'\cdot(\mathbf{z}*\mathbf{z}) & 0 & 0 \\ \mathbf{1}' & 0 & 0 & 0 & 0 \\ \mathbf{z}' & 0 & 0 & 0 & 0 \end{bmatrix} \begin{bmatrix} \mathbf{d} \\ a_0 \\ a_1 \\ \alpha_0 \\ \alpha_1 \end{bmatrix} = \begin{bmatrix} \mathbf{E}\cdot(\mathbf{w}*\mathbf{y}) \\ \mathbf{w}'\cdot\mathbf{y} \\ \mathbf{w}'\cdot(\mathbf{z}*\mathbf{y}) \\ 0 \\ 0 \end{bmatrix}.$$

We notice that with this representation of cubic splines, we no longer have the nice banded structure in the left hand matrix characteristic of the B-spline approach. However, by means of (5.36), the generalization to the bivariate case will be straightforward; this is not the case with the B-spline representation.

Following our previous route, the next step is to consider the Poisson case. The penalized deviance is

$$\begin{aligned} \Delta = 2\sum_i w_i \Bigg(& y_i \log y_i - y_i \left(\frac{1}{12}\sum_{j=1}^m d_j |x_i - z_j|^3 + a_1 + a_2 x_i \right) \\ & - y_i + \exp\left\{ \frac{1}{12}\sum_{j=1}^m d_j |x_i - z_j|^3 + a_1 + a_2 x_i \right\} \Bigg) \\ & + \frac{\lambda}{12}\sum_{j=1}^m\sum_{k=1}^m d_j d_k |z_j - z_k|^3. \end{aligned} \tag{5.42}$$

We take $\phi_1(d_1,\ldots,d_m,a_1,a_2) = 2(d_1 + \cdots + d_m)$ and $\phi_2(d_1,\ldots,d_m,a_1,a_2) = 2(d_1 z_1 + \cdots + d_m z_m)$, which does not matter since we may multiply the last two equations in (5.40) by any non-zero number. Equations (5.40) now become

$$\begin{aligned} & -\sum_{k=1}^m \tilde{w}_k \tilde{y}_k e_{kr} + \sum_{k=1}^m \tilde{w}_k \mu(z_k) e_{kr} + \lambda \sum_{k=1}^m d_k e_{kr} + \alpha_1 + \alpha_2 z_r = 0; \\ & \quad r = 1,\ldots,m; \\ & -\sum_{k=1}^m \tilde{w}_k \tilde{y}_k + \sum_{k=1}^m \tilde{w}_k \mu(z_k) = 0; \\ & -\sum_{k=1}^m \tilde{w}_k \tilde{y}_k z_k + \sum_{k=1}^m \tilde{w}_k \mu(z_k) z_k = 0; \\ & \sum_{k=1}^m d_k = 0; \\ & \sum_{k=1}^m d_k z_k = 0, \end{aligned} \tag{5.43}$$

where

$$\mu(x) = \exp\left\{\frac{1}{12}\sum_{j=1}^{m} d_j|x - z_j|^3 + a_1 + a_2 x\right\}.$$

To solve the non-linear system (5.43), we apply the Newton-Raphson method as in Sect. 5.3.2 to construct a sequence $(d_1^{(n)}, \ldots, d_m^{(n)}, a_1^{(n)}, a_2^{(n)}, \alpha_1^{(n)}, \alpha_2^{(n)})$ of estimators. Introducing the notation

$$\gamma_k^{(n)} = \exp\left\{\frac{1}{12}\sum_{j=1}^{m} d_j^{(n)}|z_k - z_j|^3 + a_1^{(n)} + a_2^{(n)} z_k\right\},$$

one finds that the resulting system of linearized equations are the same as (5.41), with $d_1, \ldots, d_m, a_1, a_2, \alpha_1, \alpha_2$ replaced by the estimators in step $n+1$, with w_k replaced by $\tilde{w}_k \gamma_k^{(n)}$ and with y_k replaced by $\tilde{y}_k/\gamma_k^{(n)} - 1 + \sum_{k=1}^{m} d_j^{(n)} e_{kj} + a_1^{(n)} + a_2^{(n)} z_k$. This is completely analogous to our findings in Sect. 5.3.2.

In the gamma case the weights are $\tilde{w}_k \tilde{y}_k/\gamma_k^{(n)}$ and the observations are $1 - \gamma_k^{(n)}/\tilde{y}_k + \sum_{k=1}^{m} d_j^{(n)} e_{kj} + a_1^{(n)} + a_2^{(n)} z_k$.

The results for natural cubic splines above are readily extended to thin plate splines. Letting $\mathbf{E}$ be the matrix with elements $e_{jk} = r_k^2(z_{1j}, z_{2j}) \log r_k^2(z_{1j}, z_{2j}) /(16\pi)$, the system of equations corresponding to (5.41) becomes

$$\begin{bmatrix}
\mathbf{E}\cdot(\mathbf{E}*\mathbf{w}) + \lambda\mathbf{E} & \mathbf{E}\cdot\mathbf{w} & \mathbf{E}\cdot(\mathbf{w}*\mathbf{z}_1) & \mathbf{E}\cdot(\mathbf{w}*\mathbf{z}_2) & \mathbf{1} & \mathbf{z}_1 & \mathbf{z}_2 \\
(\mathbf{E}\cdot\mathbf{w})' & \mathbf{w}'\cdot\mathbf{1} & \mathbf{w}'\cdot\mathbf{z}_1 & \mathbf{w}'\cdot\mathbf{z}_2 & 0 & 0 & 0 \\
(\mathbf{E}\cdot(\mathbf{w}*\mathbf{z}_1))' & \mathbf{w}'\cdot\mathbf{z}_1 & \mathbf{w}'\cdot(\mathbf{z}_1*\mathbf{z}_1) & \mathbf{w}'\cdot(\mathbf{z}_1*\mathbf{z}_2) & 0 & 0 & 0 \\
(\mathbf{E}\cdot(\mathbf{w}*\mathbf{z}_2))' & \mathbf{w}'\cdot\mathbf{z}_2 & \mathbf{w}'\cdot(\mathbf{z}_1*\mathbf{z}_2) & \mathbf{w}'\cdot(\mathbf{z}_2*\mathbf{z}_2) & 0 & 0 & 0 \\
\mathbf{1}' & 0 & 0 & 0 & 0 & 0 & 0 \\
\mathbf{z}_1' & 0 & 0 & 0 & 0 & 0 & 0 \\
\mathbf{z}_2' & 0 & 0 & 0 & 0 & 0 & 0
\end{bmatrix}
\begin{bmatrix} \mathbf{d} \\ a_0 \\ a_1 \\ a_2 \\ \alpha_0 \\ \alpha_1 \\ \alpha_2 \end{bmatrix}
= \begin{bmatrix} \mathbf{E}\cdot(\mathbf{w}*\mathbf{y}) \\ \mathbf{w}'\cdot\mathbf{y} \\ \mathbf{w}'\cdot(\mathbf{z}_1*\mathbf{y}) \\ \mathbf{w}'\cdot(\mathbf{z}_2*\mathbf{y}) \\ 0 \\ 0 \\ 0 \end{bmatrix}.$$

Apart from losing the diagonal structure originating from B-splines, the $\mathbf{E}$ matrix is sometimes nearly singular and the situation deteriorates fast as the number of distinct values m grows. A method for handling these numerical difficulties is given by Sibson and Stone [SS91].

Extending the backfitting algorithm to the situation where we have two interacting continuous variables is straightforward. For instance, suppose that instead of the second continuous variable in (5.24) we had two variables whose effect we wish to model using thin plate splines. Then $\hat{\gamma}_{2i}$ would instead look like

$$\hat{\gamma}_{2i} = \exp\left\{ \frac{1}{16\pi} \sum_{k=1}^{m} \hat{d}_k r_k^2(x_{1i}, x_{2i}) \log r_k^2(x_{1i}, x_{2i}) + \hat{a}_0 + \hat{a}_1 x_{1i} + \hat{a}_2 x_{2i} \right\}.$$

Cross-validation involves nothing new either, but can be very time-consuming due to the loss of the banded structure obtained with the B-spline approach. For this reason, a simplified version of cross-validation, called generalized cross-validation (GCV), is often used in connection with thin plate splines. For a description of GCV, see [GS94, Sect. 3.3], or [Wo06, Sect. 4.5].

5.8 Case Study: Trying GAMs in Motor Insurance

Table 5.1, available as a data set at www.math.su.se/GLMbook, contains motor TPL claim frequency data for Volvo and Saab cars, depending on car weight in tons. The purpose of the case study is to try the methods in this chapter in a simple setting. Before analyzing the data, try to imagine how claim frequencies ought to depend on car weight.

Problem 1: Use the technique described in Sect. 5.3.2 to estimate the mean claim frequencies as a function of car weight, for the "Total" column. For the smoothing parameter, use $\lambda = 10$.

Table 5.1 Motor TPL data

Car	Volvo		Saab		Total	
weight (tons)	Policy years	No of claims	Policy years	No of claims	Policy years	No of claims
1.10	23327	1759	28232	2381	51559	4140
1.15	39605	2742	34631	3110	74236	5852
1.20	25737	1219	56228	6069	81965	7288
1.25	8258	201	39595	4134	47853	4335
1.30	19078	1253	13496	1590	32574	2843
1.35	140352	9935	23994	2611	164346	12546
1.40	176091	13524	23891	2746	199982	16270
1.45	180653	13335	11300	1332	191953	14667
1.50	110373	10357	3635	553	114008	10910
1.55	133278	15140	3537	488	136816	15628
1.60	59722	7616	4918	602	64640	8218
1.65	33219	4455	4188	503	37407	4958
1.70	15990	2367	2103	309	18094	2676

Problem 2: Treating the variables *car weight* and *car brand* (Volvo/Saab) as non-interacting, estimate the relativities for these two rating factors, using the method in Sect. 5.4.2. Choose the smoothing parameter λ for car weight subjectively using a partial residual plot.

Problem 3: For the data in Problem 1, use the technique described in Sect. 5.5 to choose the smoothing parameter λ.

N.B. The data consists of a certain subset of a large motor portfolio and no conclusions concerning Volvo and Saab owners in general should be drawn.

Exercises

For the exercises in this chapter that require programming, some software that handles matrices, like Matlab or SAS/IML, is helpful.

5.1 (Section 5.2) Starting from (B.9), find the unique natural cubic spline interpolating the following points in the (x, y)-plane:

x	0.0	0.2	0.5	0.7	0.8	1.2	1.3	1.5	1.9	2.5	2.7	3.1	3.3	3.9
y	1.3	1.7	2.3	1.9	1.6	2.1	2.4	2.8	1.6	1.2	0.9	1.3	1.9	1.2

Using (B.6), make a plot of the points and the interpolating spline.

5.2 (Section 5.2) For the set of knots $0.0, 0.1, 0.2, \ldots, 1.0$, produce a graph similar to Fig. 5.4 of the B-spline basis, using the basic recursion formula (B.17) in Appendix B.2. Also, using Propositions B.1 and B.2, produce the corresponding graphs of the first and second derivatives.

5.3 (Section 5.3) Carry out the derivation of (5.16) in detail.

5.4 (Section 5.3) Changing the scale for a continuous rating variable, for instance by measuring vehicle weight in tons instead of kilograms, affects the penalized deviance. This has led to the suggestion that one should use

$$\Delta(f) = D(y, \mu) + \lambda^3 \int_a^b (f''(x))^2\, dx, \tag{5.44}$$

instead of (5.4). Assume that we have a single rating variable x_i and that f in (5.44) is a cubic spline. Using the results of Appendix B.2, show that if we replace x_i by cx_i and λ by $c\lambda$, the penalized deviance (5.44) remains the same.

5.5 (Section 5.7) Prove Lemma B.6.

Appendix A
Some Results from Probability and Statistics

Here we have collected some results from probability and statistics that are used in the book.

A.1 The Gamma Function

The gamma function is used in several probability distribution functions. It is defined for real $p > 0$ as

$$\Gamma(p) = \int_0^\infty x^{p-1} \mathrm{e}^{-x}\, dx.$$

Some important properties of this function are

(a) $\Gamma(p) = (p-1)\Gamma(p-1)$;
(b) $\Gamma(1) = 1$;
(c) For positive integers n we have $\Gamma(n) = (n-1)!$;
(d) $\Gamma(\frac{1}{2}) = \sqrt{\pi}$.

Proofs of these properties may be found in textbooks on probability theory, e.g. [Ro02, Chap. 5.6.1].

A.2 Conditional Expectation

First we repeat two well-known and useful results, see e.g. Theorem 2.1. and Corollary 2.3.1 in [Gu95].

Lemma A.1 *For any random variables X and Y with $E(|Y|) < \infty$ we have*

$$E(Y) = E[E(Y|X)]. \tag{A.1}$$

E. Ohlsson, B. Johansson, *Non-Life Insurance Pricing with Generalized Linear Models*, EAA Lecture Notes,
DOI 10.1007/978-3-642-10791-7,

Lemma A.2 *For any random variables X and Y with $E(Y^2) < \infty$ we have*

$$\mathrm{Var}(Y) = E[\mathrm{Var}(Y|X)] + \mathrm{Var}[E(Y|X)]. \tag{A.2}$$

We now generalize the last result to covariances.

Lemma A.3 *Let X and Y be random variables with $E(X^2) < \infty$ and $E(Y^2) < \infty$, respectively, and Z a random vector.*

(a) *Then*

$$\mathrm{Cov}(X, Y) = E[\mathrm{Cov}(X, Y|Z)] + \mathrm{Cov}[E(X|Z), E(Y|Z)].$$

(b) *When X is a function of Z this specializes to*

$$\mathrm{Cov}(X, Y) = \mathrm{Cov}[X, E(Y|Z)].$$

Note that Lemma A.2 is the special case of the result in Lemma A.3(a) where $X = Y$ and Z is scalar.

Proof

$$\begin{aligned}
\mathrm{Cov}(X, Y) &= E(XY) - E(X)E(Y) \\
&= E(XY - E[X|Z]E[Y|Z]) \\
&\quad + E(E[X|Z]E[Y|Z] - E[X]E[Y]) \\
&= E(E[XY|Z] - E[X|Z]E[Y|Z]) \\
&\quad + E(E[X|Z]E[Y|Z] - E[E(X|Z)]E[E(Y|Z)]) \\
&= E[\mathrm{Cov}(X, Y|Z)] + \mathrm{Cov}[E(X|Z), E(Y|Z)].
\end{aligned}$$

□

A.3 The Law of Total Probability

Let $Y = 1$ if the event A occurs, and else 0. Note that $E(Y) = P(A)$.

Lemma A.4

(a) *Law of total probability for a discrete random variable X.*

$$P(A) = \sum_k P(A|X = x_k)p_X(x_k).$$

(b) *Law of total probability for a continuous random variable X.*

$$P(A) = \int P(A|X = x)f_X(x)\,dx.$$

These results are found directly by using Lemma A.1 on Y.

A.4 Bayes' Theorem

Let (X, Y) be a two-dimensional random variable and let $f_{Y|X}(y|x)$ denote the frequency function for Y given $X = x$, i.e., the probability mass function in the discrete case or the probability density function in the continuous case. Upon observing $X = x$ we may "update" a given *prior distribution* $f_Y(y)$ to a *posterior distribution* $f_{Y|X}(y|x)$, by using *Bayes' theorem.*

Lemma A.5 (Bayes' theorem) *For random variables X and Y we have*

$$f_{Y|X}(y|x) = \frac{f_{X|Y}(x|y)\, f_Y(y)}{\int f_{X|Y}(x|t)\, f_Y(t)\, dt}.$$

In case Y is discrete, the integral "$\int dt$" should be read as a sum, "$\sum_t$".

Proof The case when X and Y are both discrete is well known, see, e.g., equation (3.4) on p. 6 of [Gu95]. When both X and Y are continuous we use the definition of a conditional density to find

$$f_{Y|X}(y|x) \doteq \frac{f_{X,Y}(x, y)}{f_X(x)} = \frac{f_{X|Y}(x|y)\, f_Y(y)}{\int f_{X,Y}(x, t)\, dt} = \frac{f_{X|Y}(x|y)\, f_Y(y)}{\int f_{X|Y}(x|t)\, f_Y(t)\, dt}.$$

If X is discrete and Y continuous, we work with the conditional distribution function

$$F_{Y|X}(y|x) = \frac{P(Y \le y, X = x)}{P(X = x)}. \tag{A.3}$$

By using Lemma A.4(b)

$$\begin{aligned} P(Y \le y, X = x) &= \int_{-\infty}^{+\infty} P(Y \le y, X = x | Y = t)\, f_Y(t)\, dt \\ &= \int_{-\infty}^{y} P(X = x | Y = t)\, f_Y(t)\, dt. \end{aligned}$$

We substitute this in (A.3) and differentiate with respect to y and get

$$f_{Y|X}(y|x) = \frac{f_{X|Y}(x|y)\, f_Y(y)}{P(X = x)}.$$

Another application of Lemma A.4(b), this time on the denominator of (A.3) yields the result.

The proof for the case with X continuous and Y discrete is omitted here, since that result is not used in this book. □

A.5 Unbiased Estimation of Weighted Variances

Lemma A.6 *Let $X_1, X_2, \ldots, X_n$ be a sequence of uncorrelated random variables with common mean μ and with variance proportional to known weights w_k, $\mathrm{Var}(X_k) = \sigma^2/w_k$; $k = 1, 2, \ldots, n$. With $w_\cdot = \sum_k w_k$ we define*

$$\overline{X}^{(w)} = \frac{1}{w_\cdot} \sum_k w_k X_k,$$

$$s^{2(w)} = \frac{1}{n-1} \sum_k w_k (X_k - \overline{X}^{(w)})^2.$$

These statistics are then unbiased estimators of the model parameters, i.e.,

$$E(\overline{X}^{(w)}) = \mu, \qquad E(s^{2(w)}) = \sigma^2.$$

Proof It is trivial that $\overline{X}^{(w)}$ is unbiased and we restrict the proof to $s^{2(w)}$.

$$\mathrm{Var}(X_k - \overline{X}^{(w)}) = \mathrm{Var}\left(X_k \left(1 - \frac{w_k}{w_\cdot}\right) - \sum_{i \neq k} \frac{w_i}{w_\cdot} X_i \right)$$

$$= \left(1 - \frac{w_k}{w_\cdot}\right)^2 \frac{\sigma^2}{w_k} + \sum_{i \neq k} \frac{w_i}{w_\cdot^2} \sigma^2 = \frac{\sigma^2}{w_k}\left(1 - \frac{w_k}{w_\cdot}\right), \tag{A.4}$$

so that

$$E\left[\sum_{k=1}^{n} w_k (X_k - \overline{X}^{(w)})^2\right] = \sum_{k=1}^{n} w_k \, \mathrm{Var}(X_k - \overline{X}^{(w)}) = \sigma^2 (n-1).$$

□

Appendix B
Some Results on Splines

B.1 Cubic Splines

In this section we show why cubic splines play a central role in connection with the penalized deviance (5.4). These results will also provide guidance for constructing a bivariate generalization of univariate splines. The starting point is the following lemma.

Lemma B.1 *Consider a set of points $u_1 < \cdots < u_m$ and real numbers $y_1, \ldots, y_m$, and let $\mathcal{I}$ denote the set of twice continuously differentiable functions f such that $f(u_k) = y_k$, $k = 1, \ldots, m$. Assume that $s \in \mathcal{I}$ has the property that for any twice continuously differentiable function h with $h(u_k) = 0$; $k = 1, \ldots, m$,*

$$\int_a^b s''(x)h''(x)\,dx = 0, \quad a < u_1,\ b > u_m. \tag{B.1}$$

Then for any $f \in \mathcal{I}$,

$$\int_a^b (s''(x))^2 dx \le \int_a^b (f''(x))^2 dx, \quad a < u_1,\ b > u_m.$$

Proof Writing $h(x) = f(x) - s(x)$, we have $h(u_k) = 0$, $k = 1, \ldots, m$, and so

$$\begin{aligned}\int_a^b (f''(x))^2\,dx &= \int_a^b \left((s''(x))^2 + 2s''(x)h''(x) + (h''(x))^2\right) dx \\ &= \int_a^b (s''(x))^2\,dx + \int_a^b (h''(x))^2\,dx \ge \int_a^b (s''(x))^2\,dx. \qquad \square\end{aligned}$$

We next look more closely at the property (B.1). We shall assume that h is twice continuously differentiable, but not necessarily satisfying $h(u_k) = 0$, $k = 1, \ldots, m$.

E. Ohlsson, B. Johansson, *Non-Life Insurance Pricing with Generalized Linear Models*, EAA Lecture Notes,
DOI 10.1007/978-3-642-10791-7, © Springer-Verlag Berlin Heidelberg 2010

To ease the notation, we write $u_0 = a$ and $u_{m+1} = b$. We then have

$$\int_a^b s''(x)h''(x)\,dx = \lim_{\varepsilon\to 0}\sum_{k=0}^{m}\int_{u_k+\varepsilon}^{u_{k+1}-\varepsilon} s''(x)h''(x)\,dx.$$

Assuming that s is four times continuously differentiable for $x \neq u_1,\ldots,u_m$, integrating by parts twice gives

$$\begin{aligned}\int_{u_k+\varepsilon}^{u_{k+1}-\varepsilon} s''(x)h''(x)\,dx &= s''(u_{k+1}-\varepsilon)h'(u_{k+1}-\varepsilon) - s''(u_k+\varepsilon)h'(u_k+\varepsilon)\\ &\quad - s'''(u_{k+1}-\varepsilon)h(u_{k+1}-\varepsilon) + s'''(u_k+\varepsilon)h(u_k+\varepsilon)\\ &\quad + \int_{u_k+\varepsilon}^{u_{k+1}-\varepsilon} \frac{d^4}{dx^4}s(x)h(x)\,dx.\end{aligned}$$

We now observe that if $d^4s(x)/dx^4 = 0$ for $x \neq u_1,\ldots,u_m$, then the final integral in the above expression disappears. Assuming this (implying that $s'''(x)$ is constant between the knots), and letting $\varepsilon \to 0$, we get

$$\begin{aligned}\int_a^b s''(x)h''(x)\,dx &= \sum_{k=0}^{m}[s''(u_{k+1})h'(u_{k+1}) - s''(u_k)h'(u_k)]\\ &\quad - \sum_{k=0}^{m}[s'''(u_{k+1}-)h(u_{k+1}) - s'''(u_k+)h(u_k)]\\ &= s''(u_{m+1})h'(u_{m+1}) - s''(u_0)h'(u_0)\\ &\quad - \sum_{k=1}^{m}(s'''(u_k+) - s'''(u_k-))h(u_k).\end{aligned}$$

We see that if $s''(x)$ is identically zero for $x \leq u_1$ and $x \geq u_m$ then (B.1) will hold when $h(u_k) = 0$ for $k = 1,\ldots,m$. Let us summarize our calculations in the following lemma.

Lemma B.2 *Assume that s is four times continuously differentiable, except at the points $u_1,\ldots,u_m$, where it is twice continuously differentiable. Furthermore, assume that s satisfies the following conditions:*

(i) $\frac{d^4}{dx^4}s(x) = 0$ *for* $x \neq u_1,\ldots,u_m$;
(ii) $s''(x) = 0$ *for* $x \leq u_1$ *and* $x \geq u_m$.

Then for any twice continuously differentiable function h and for any $a < u_1$, $b > u_m$,

$$\int_a^b s''(x)h''(x)\,dx = \sum_{k=1}^{m} d_k h(u_k), \tag{B.2}$$

where $d_k = s'''(u_k+) - s'''(u_k-)$.

From (i) it follows that s must be a cubic polynomial on each of the intervals (u_k, u_{k+1}) and if it is twice continuously differentiable, it must be a cubic spline. Furthermore, the condition (ii) says that it is natural. Starting from the conditions (i) and (ii), we shall now construct an explicit expression for a natural cubic spline.

So we start from the fact that s is twice continuously differentiable and that it satisfies (i) and (ii) in Lemma B.2. From (i) it follows that $s'''(x)$ must be piecewise constant between any two points u_k and u_{k+1}. Furthermore, by (ii), $s'''(x)$ must be identically zero for $x < u_1$ and $x > u_m$. Therefore, if $x \neq u_1, \ldots, u_m$,

$$s'''(x) = \sum_{k=1}^{m} d_k I_{\{u_k<x\}}, \tag{B.3}$$

with d_k as in Lemma B.2. We note that

$$\sum_{k=1}^{m} d_k = 0, \qquad \sum_{k=1}^{m} d_k u_k = 0. \tag{B.4}$$

The first equality follows from the fact that $\sum_k d_k = s'''(b) - s'''(a) = 0$, where $a < u_1$, $b > u_m$. The second holds because $\sum_k d_k u_k = -\int_a^b s'''(x)\,dx = s''(a) - s''(b) = 0$.

We now derive an expression for $s''(x)$, starting from (B.3). The primitive function of $t \mapsto d_k I_{\{u_k<x\}}$ is, if $x \neq u_k$, $x \mapsto b_k I_{\{u_k<x\}} + (d_k x + b_k) I_{\{x>u_k\}}$, for some constants b_k and c_k. Thus, if $x \neq u_1, \ldots, u_m$,

$$s''(x) = \sum_{k=1}^{m} \big(b_k I_{\{u_k<x\}} + (d_k x + c_k) I_{\{x>u_k\}}\big) + a,$$

for some constant a. By letting $x \to u_k$ from above and below, we have, due to the continuity of $s''(x)$, that $b_k = d_k u_k + c_k$. From this we get

$$s''(x) = \sum_{k=1}^{m} \big(d_k u_k I_{\{u_k<x\}} + d_k x I_{\{x>u_k\}}\big) + b,$$

where $b = \sum_{k=1}^{m} c_k + a$. Since for $x > u_m$, $s''(x) = \sum_{k=1}^{m} d_k x + b$, we conclude from (B.4) that $b = 0$. Again, using (B.4) we have

$$\begin{aligned} s''(x) &= \sum_{k=1}^{m} d_k u_k I_{\{u_k<x\}} + \sum_{k=1}^{m} d_k x I_{\{x>u_k\}} \\ &= \sum_{k=1}^{m} d_k u_k I_{\{u_k<x\}} + x \left(\sum_{k=1}^{m} d_k - \sum_{k=1}^{m} d_k I_{\{x<u_k\}} \right) \\ &= \sum_{k=1}^{m} d_k (u_k - x) I_{\{x<u_k\}}. \end{aligned}$$

On the other hand,

$$
\begin{aligned}
s''(x) &= \sum_{k=1}^{m} d_k u_k I_{\{u_k<x\}} + \sum_{k=1}^{m} d_k x I_{\{x>u_k\}} \\
&= \sum_{k=1}^{m} d_k u_k - \sum_{k=1}^{m} d_k u_k I_{\{x>u_k\}} + \sum_{k=1}^{m} d_k x I_{\{x>u_k\}} \\
&= \sum_{k=1}^{m} d_k (x-u_k) I_{\{x>u_k\}}.
\end{aligned}
$$

We thus have

$$
\begin{aligned}
2s''(x) &= s''(x) + s''(x) \\
&= \sum_{k=1}^{m} d_k (u_k - x) I_{\{x<u_k\}} + \sum_{k=1}^{m} d_k (x-u_k) I_{\{x>u_k\}} \\
&= \sum_{k=1}^{m} d_k |x-u_k|,
\end{aligned}
$$

and so

$$
s''(x) = \frac{1}{2} \sum_{k=1}^{m} d_k |x-u_k|. \tag{B.5}
$$

From (B.5) the following lemma is easily shown, using (B.4) again.

Lemma B.3 *Assume that the twice continuously differentiable function s satisfies* (i) *and* (ii) *in Lemma* B.2. *Then*

$$
s(x) = \frac{1}{12} \sum_{k=1}^{m} d_k |x-u_k|^3 + a_0 + a_1 x, \tag{B.6}
$$

for some constants a_0 and a_1.

This is an explicit representation of a natural cubic spline: it is twice continuously differentiable and between the knots it is a cubic polynomial; furthermore, the condition (B.4) implies that it is natural. The next lemma shows that with this representation, a simple expression for the integrated squared second derivative can be easily derived.

Lemma B.4 *With $s(x)$ as in* (B.6),

$$
\int_a^b (s''(x))^2\, dx = \frac{1}{12} \sum_{j=1}^{m} \sum_{k=1}^{m} d_j d_k |u_j - u_k|^3. \tag{B.7}
$$

Proof Taking $h = s$ in (B.2), we get

$$\int_a^b (s''(x))^2\,dx = \sum_{j=1}^m d_j s(u_j)$$

$$= \frac{1}{12}\sum_{j=1}^m d_j \sum_{k=1}^m d_k |u_j - u_k|^3 + a_0 \sum_{j=1}^m d_j + a_1 \sum_{j=1}^m d_j u_j.$$

Thus (B.7) follows from (B.4). □

We are now ready to prove the following basic result.

Theorem B.1 *For any $u_1, \ldots, u_m$ with $u_1 < \cdots < u_m$ and any real numbers $y_1, \ldots, y_m$ there exists a unique natural cubic spline $s(x)$, such that $s(u_j) = y_j$; $j = 1, \ldots, m$.*

Proof Put $e_{jk} := |u_j - u_k|^3/12$. By Lemma B.3 we may represent natural cubic splines as in (B.6). The conditions $s(u_1) = y_1, \ldots, s(u_m) = y_m$ then become

$$\begin{aligned} e_{11}d_1 + e_{12}d_2 + \cdots + e_{1m}d_m + a_0 + a_1 u_1 &= y_1, \\ e_{21}d_1 + e_{22}d_2 + \cdots + e_{2m}d_m + a_0 + a_1 u_2 &= y_2, \\ \vdots \quad\quad \vdots \\ e_{m1}d_1 + e_{m2}d_2 + \cdots + e_{mm}d_m + a_0 + a_1 u_m &= y_m. \end{aligned} \tag{B.8}$$

These are m equations with $m + 2$ unknowns, but if we add those in (B.4), we get $m + 2$ equations. We introduce the matrix

$$\mathbf{E} = \begin{bmatrix} e_{11} & e_{12} & \cdots & e_{1m} \\ e_{21} & e_{22} & \cdots & e_{2m} \\ \vdots & \vdots & & \vdots \\ e_{m1} & e_{m2} & \cdots & e_{mm} \end{bmatrix},$$

and the vectors

$$\mathbf{d} = \begin{bmatrix} d_1 \\ d_2 \\ \vdots \\ d_m \end{bmatrix}, \quad \mathbf{1} = \begin{bmatrix} 1 \\ 1 \\ \vdots \\ 1 \end{bmatrix}, \quad \mathbf{u} = \begin{bmatrix} u_1 \\ u_2 \\ \vdots \\ u_m \end{bmatrix}, \quad \mathbf{y} = \begin{bmatrix} y_1 \\ y_2 \\ \vdots \\ y_m \end{bmatrix}.$$

We may then write (B.8) plus (B.4) using block matrices as

$$\begin{bmatrix} \mathbf{E} & \mathbf{1} & \mathbf{u} \\ \mathbf{1}' & 0 & 0 \\ \mathbf{u}' & 0 & 0 \end{bmatrix} \begin{bmatrix} \mathbf{d} \\ a_0 \\ a_1 \end{bmatrix} = \begin{bmatrix} \mathbf{y} \\ 0 \\ 0 \end{bmatrix}. \tag{B.9}$$

If we can show that these equations have a unique solution, then the natural cubic spline (B.6) with parameters $d_1, \ldots, d_m, a_0, a_1$ corresponding to the solution will be a (unique) interpolating natural cubic spline. It suffices to show that the matrix on the left in (B.9) has full rank. To do this, we show that (using $\mathbf{0}$ to denote a vector of m zeroes)

$$\begin{bmatrix} \mathbf{E} & \mathbf{1} & \mathbf{u} \\ \mathbf{1}' & 0 & 0 \\ \mathbf{u}' & 0 & 0 \end{bmatrix} \begin{bmatrix} \mathbf{d} \\ a_0 \\ a_1 \end{bmatrix} = \begin{bmatrix} \mathbf{0} \\ 0 \\ 0 \end{bmatrix} \tag{B.10}$$

implies

$$\begin{bmatrix} \mathbf{d} \\ a_0 \\ a_1 \end{bmatrix} = \begin{bmatrix} \mathbf{0} \\ 0 \\ 0 \end{bmatrix}. \tag{B.11}$$

If (B.10) holds, we have

$$\begin{bmatrix} \mathbf{d}' & a_0 & a_1 \end{bmatrix} \begin{bmatrix} \mathbf{E} & \mathbf{1} & \mathbf{u} \\ \mathbf{1}' & 0 & 0 \\ \mathbf{u}' & 0 & 0 \end{bmatrix} \begin{bmatrix} \mathbf{d} \\ a_0 \\ a_1 \end{bmatrix} = 0. \tag{B.12}$$

But if $s(x)$ is the natural cubic spline with the parameters $d_1, \ldots, d_m, a_0, a_1$ in (B.12), then by Lemma B.4 and (B.4),

$$\begin{aligned} &\begin{bmatrix} \mathbf{d}' & a_0 & a_1 \end{bmatrix} \begin{bmatrix} \mathbf{E} & \mathbf{1} & \mathbf{u} \\ \mathbf{1}' & 0 & 0 \\ \mathbf{u}' & 0 & 0 \end{bmatrix} \begin{bmatrix} \mathbf{d} \\ a_0 \\ a_1 \end{bmatrix} \\ &= \sum_{j=1}^{m} \sum_{k=1}^{m} d_j d_k e_{jk} + 2a_0 \sum_{j=1}^{m} d_j + 2a_1 \sum_{j=1}^{m} d_j u_j = \int \left(s''(x)\right)^2 dx. \end{aligned} \tag{B.13}$$

But (B.12) and (B.13) put together implies that $s(x)$ must be linear, which means that the jumps in the third derivatives, i.e. the d_j, are all zero. Furthermore, using this in (B.8), we get

$$\begin{aligned} a_0 + a_1 u_1 &= 0, \\ a_0 + a_1 u_2 &= 0, \\ &\vdots \\ a_0 + a_1 u_m &= 0. \end{aligned}$$

Since the points $u_1, \ldots, u_m$ are distinct, this implies that $a_1 = a_2 = 0$. □

Finally, we have the result stating that among all twice continuously differentiable functions with given values at certain points, the interpolating natural cubic spline minimizes the integrated squared second derivative.

Theorem B.2 *Let $u_1 < \cdots < u_m$, let $f(\cdot)$ be any twice continuously differentiable function and let $s(\cdot)$ be the natural cubic spline satisfying $s(u_j) = f(u_j)$, $j = 1, \ldots, m$. Then, for any $a \le u_1$ and $b \ge u_m$,*

$$\int_a^b (s''(x))^2\,dx \le \int_a^b (f''(x))^2\,dx. \tag{B.14}$$

Proof It follows from Lemma B.2 that s satisfies (B.1) for any twice differentiable h with $h(u_k) = 0$, $k = 1, \ldots, m$. Thus the result follows from Lemma B.1. □

B.2 B-splines

It is immediate that the set of cubic splines with a given set of knots $u_1, \ldots, u_m$ forms a linear space, and the same is true for quadratic and linear splines. What is the dimension of these linear spaces? To begin with, a linear spline consists of $m-1$ linear polynomials, each having two parameters. The continuity restriction at the internal knots yields $m-2$ linear conditions involving the parameters. Thus the dimension of the set of linear splines should be $2(m-1)-(m-2) = m$. Similarly, the dimension of the set of quadratic and cubic splines would be $3(m-1)-2(m-2) = m+1$ and $4(m-1)-3(m-2) = m+2$, respectively. We will soon verify that this is the case.

We begin by defining a base for step functions with jumps at the knots. For $k = 1, \ldots, m-2$, put

$$B_{0,k}(x) = \begin{cases} 1, & u_k \le x < u_{k+1}; \\ 0, & \text{otherwise}, \end{cases} \tag{B.15}$$

and furthermore

$$B_{0,m-1}(x) = \begin{cases} 1, & u_{m-1} \le x \le u_m; \\ 0, & \text{otherwise}. \end{cases} \tag{B.16}$$

It is a matter of convention whether we choose to define these basis functions as left or right continuous and it does not matter for what follows. Obviously, any left continuous step function may be written as a linear combination of the functions $B_{0,k}$.

For $j \ge 0$, we define the *B-splines* recursively by

$$B_{j+1,k}(x) = \frac{x - u_{k-j-1}}{u_k - u_{k-j-1}} B_{j,k-1}(x) + \frac{u_{k+1} - x}{u_{k+1} - u_{k-j}} B_{j,k}(x); \quad k = 1, \ldots, m+j. \tag{B.17}$$

In (B.17), usually referred to as *de Boor's recursion formula*, $B_{j,k}(x)$ is defined to be zero if $k \le 0$ or if $k \ge m+j$. Furthermore, we define $u_k = u_1$ for $k \le 0$ and $u_k = u_m$ for $k \ge m+1$. It is easy to see that $B_{j,k}(x)$ is positive on (u_{k-j}, u_{k+1}) and zero otherwise. Examples of linear, quadratic and cubic B-splines are shown in

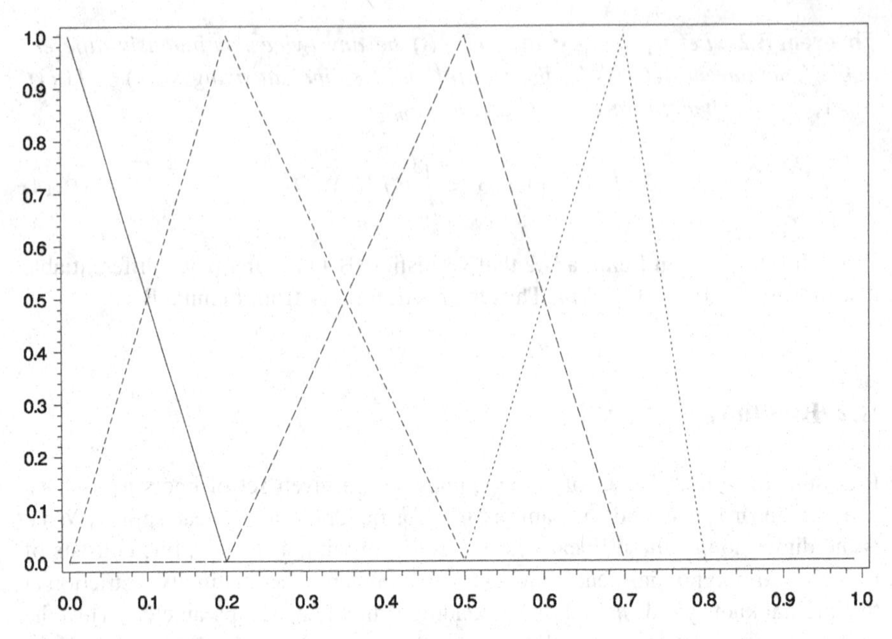

Fig. B.1 B-splines of order 1

Figs. B.1, B.2 and B.3. That the B-splines are continuous is immediate from the definition. We will now go on to show that they also satisfy the required differentiability conditions. From the next proposition it follows that for $j \geq 2$, $B_{j,k}$ is continuously differentiable.

Proposition B.1 *For $j \geq 0$ and $x \neq u_1, \ldots, u_m$,*

$$B'_{j+1,k}(x) = \frac{j+1}{u_k - u_{k-j-1}} B_{j,k-1}(x) - \frac{j+1}{u_{k+1} - u_{k-j}} B_{j,k}(x). \tag{B.18}$$

Note: For $j \geq 1$, it follows from the continuity of $B_{j,k-1}$ and $B_{j,k}$ that (B.18) also holds at the knots.

Proof For $j = 0$, (B.18) is easily verified directly upon differentiating in (B.17). Now assuming that (B.18) holds for j, we shall prove that it also holds for $j+1$. Differentiating (B.17), we get

$$\begin{aligned} B'_{j+1,k}(x) = & \frac{1}{u_k - u_{k-j-1}} B_{j,k-1}(x) + \frac{x - u_{k-j-1}}{u_k - u_{k-j-1}} B'_{j,k-1}(x) \\ & - \frac{1}{u_{k+1} - u_{k-j}} B_{j,k}(x) + \frac{u_{k+1} - x}{u_{k+1} - u_{k-j}} B'_{j,k}(x). \end{aligned} \tag{B.19}$$

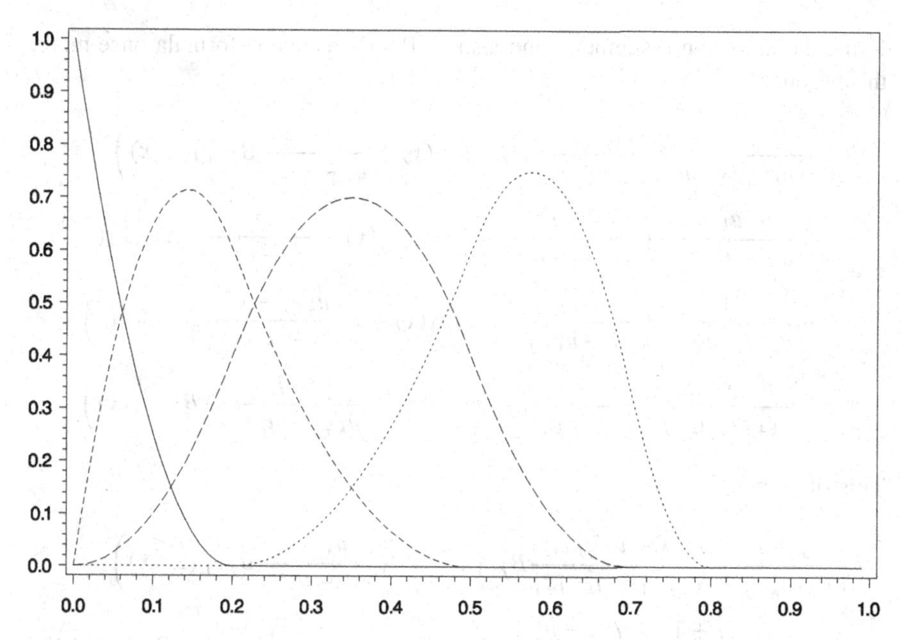

Fig. B.2 B-splines of order 2

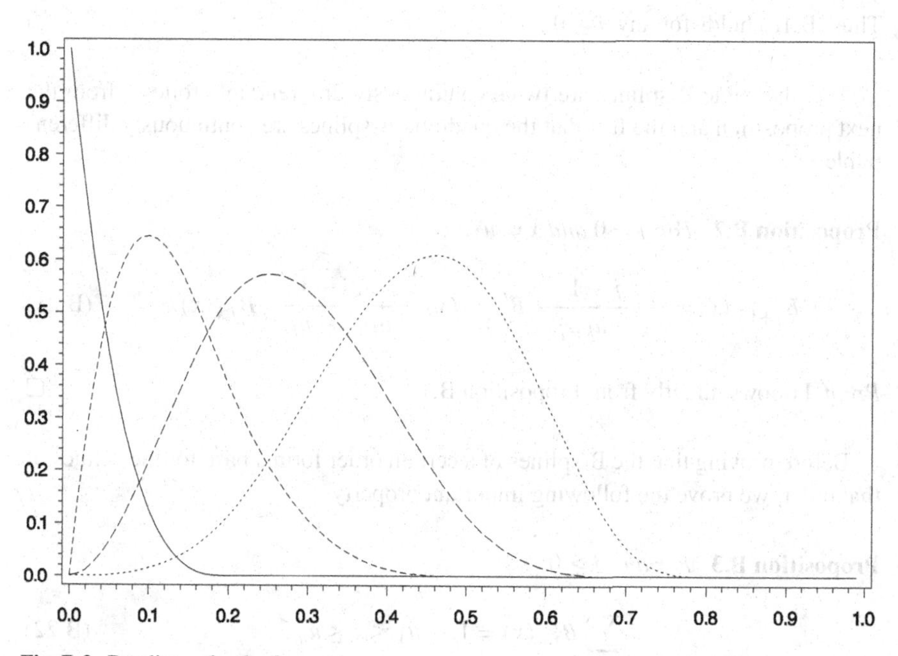

Fig. B.3 B-splines of order 3

Using the induction assumption and also de Boor's recursion formula once more, this becomes

$$\frac{1}{u_k-u_{k-j-1}}\left(\frac{x-u_{k-j-1}}{u_{k-1}-u_{k-j-1}}B_{j-1,k-2}(x)+\frac{u_k-x}{u_k-u_{k-j}}B_{j-1,k-1}(x)\right)$$
$$+\frac{x-u_{k-j-1}}{u_k-u_{k-j-1}}\left(\frac{j}{u_{k-1}-u_{k-j-1}}B_{j-1,k-2}(x)-\frac{j}{u_k-u_{k-j}}B_{j-1,k-1}(x)\right)$$
$$-\frac{1}{u_{k+1}-u_{k-j}}\left(\frac{x-u_{k-j}}{u_k-u_{k-j}}B_{j-1,k-1}(x)+\frac{u_{k+1}-x}{u_{k+1}-u_{k-j+1}}B_{j-1,k}(x)\right)$$
$$+\frac{u_{k+1}-x}{u_{k+1}-u_{k-j}}\left(\frac{j}{u_k-u_{k-j}}B_{j-1,k-1}(x)-\frac{j}{u_{k+1}-u_{k-j+1}}B_{j-1,k}(x)\right).$$

This simplifies to

$$\frac{j+1}{u_k-u_{k-j-1}}\left(\frac{x-u_{k-j-1}}{u_{k-1}-u_{k-j-1}}B_{j-1,k-2}(x)+\frac{u_k-x}{u_k-u_{k-j}}B_{j-1,k-1}(x)\right)$$
$$-\frac{j+1}{u_{k+1}-u_{k-j}}\left(\frac{x-u_{k-j}}{u_k-u_{k-j}}B_{j-1,k-1}(x)+\frac{u_{k+1}-x}{u_{k+1}-u_{k-j+1}}B_{j-1,k}(x)\right)$$
$$=\frac{j+1}{u_k-u_{k-j-1}}B_{j,k-1}(x)-\frac{j+1}{u_{k+1}-u_{k-j}}B_{j,k}(x). \tag{B.20}$$

Thus (B.18) holds for any $j \ge 0$. □

That the cubic B-splines are twice continuously differentiable follows from the next proposition and the fact that the quadratic B-splines are continuously differentiable.

Proposition B.2 *For $j \ge 0$ and $x \ne u_1, \dots, u_m$,*

$$B''_{j+1,k}(x)=\frac{j+1}{u_k-u_{k-j-1}}B'_{j,k-1}(x)-\frac{j+1}{u_{k+1}-u_{k-j}}B'_{j,k}(x). \tag{B.21}$$

Proof Follows directly from Proposition B.1. □

Before proving that the B-splines of a certain order form a base for the splines of that order, we prove the following important property.

Proposition B.3 *For any $j \ge 0$,*

$$\sum_k B_{j,k}(x)=1, \quad u_1 \le x \le u_m. \tag{B.22}$$

Proof Again the proof is by induction. For $j = 0$ the statement is trivially true. Assuming that it holds for a certain j, de Boor's recursion formula gives

$$\begin{aligned}\sum_k B_{j+1,k}(x) &= \sum_k \frac{x - u_{k-j-1}}{u_k - u_{k-j-1}} B_{j,k-1}(x) + \sum_k \frac{u_{k+1} - x}{u_{k+1} - u_{k-j}} B_{j,k}(x) \\ &= \sum_k \frac{x - u_{k-j}}{u_{k+1} - u_{k-j}} B_{j,k}(x) + \sum_k \frac{u_{k+1} - x}{u_{k+1} - u_{k-j}} B_{j,k}(x) \\ &= \sum_k B_{j,k}(x) = 1. \end{aligned} \tag{B.23}$$

□

The proof of the basis theorem uses the following lemma.

Lemma B.5 *Suppose $s(x)$ is a linear combination of B-splines of order $j+1$:*

$$s(x) = \sum_k \alpha_k B_{j+1,k}(x).$$

Then

$$s'(x) = \sum_k (j+1) \frac{\alpha_{k+1} - \alpha_k}{u_{k+1} - u_{k-j}} B_{j,k}(x).$$

Proof This is a simple consequence of Proposition B.1. □

Theorem B.3 *For a given set of knots, a spline s of order j may be written as*

$$s(x) = \sum_{k=1}^{m+j-1} \beta_k B_{j,k},$$

for unique constants $\beta_1, \ldots, \beta_{m+j-1}$.

Proof For $j = 1$, we may write

$$s(x) = \sum_{k=1}^{m} \beta_k B_{1,k}(x), \tag{B.24}$$

with $\beta_k = s(u_k)$. The two sides of (B.24) coincide since they are both linear splines and agree at the knots. Also, the representation is unique.

Assume now that the result holds for some $j \geq 1$ and consider a spline s of order $j+1$. Since s' is a spline of order j, it may by assumption be represented as

$$s'(x) = \sum_k \beta_k B_{j,k}(x).$$

Now let $\alpha_1 = 0$ and define $\alpha_2, \ldots, \alpha_{m+j-1}$ recursively by

$$\alpha_{j+1} = \alpha_j + \beta_j \frac{u_{k+1} - u_{k-j}}{j+1}.$$

If we define

$$S(x) = \sum_{k=1}^{m+j-1} \alpha_k B_{j+1,k}(x),$$

then by Lemma B.5 we have $S'(x) = s'(x)$ and $S(u_1) = 0$. Thus, using Proposition B.3,

$$\begin{aligned} s(x) &= s(u_1) + \int_{u_1}^{x} s'(y)\,dy = s(u_1) + S(x) \\ &= \sum_{k=1}^{m+j-1} s(u_1) B_{j+1,k}(x) + \sum_{k=1}^{m+j-1} \alpha_k B_{j+1,k}(x) \\ &= \sum_{k=1}^{m+j-1} (s(u_1) + \alpha_k) B_{j+1,k}(x). \end{aligned}$$

Given that $\alpha_1 = 0$, the coefficients $\alpha_2, \ldots, \alpha_{m+j-1}$ are determined by the values of $\beta_1, \ldots, \beta_{m+j-2}$. If α_1 was set to anything else, it would still hold that $S(x) - S(u_1)$ would be determined by $\beta_1, \ldots, \beta_{m+j-2}$. Thus uniqueness for splines of degree j implies uniqueness for degree $j+1$. □

Finally, we turn to computation of the Ω matrix, utilizing the recursion formulae derived above. Put

$$\begin{aligned} \Omega_{1k\ell} &= \int B_{1k}(x) B_{1\ell}(x)\,dx, \\ \Omega_{2k\ell} &= \int B'_{2k}(x) B'_{2\ell}(x)\,dx, \\ \Omega_{3k\ell} &= \int B''_{3k}(x) B''_{3\ell}(x)\,dx. \end{aligned}$$

Let $u_1, \ldots, u_m$ denote the knots. Recall the conventions $u_k = u_1$ for $k \leq 0$ and $u_k = u_m$ for $k \geq m+1$. With these in mind, put

$$\begin{aligned} a_{2k} &= \frac{2}{u_{k+1} - u_{k-1}}; \quad k = 1, \ldots, m; \\ a_{3k} &= \frac{3}{u_{k+1} - u_{k-2}}; \quad k = 1, \ldots, m+1. \end{aligned}$$

The following proposition describes how to compute the Ω matrix. Note that, due to the convention $B_{jk} = 0$ for $k \leq 0$ and $k \geq m + j$, certain $\Omega_{jk\ell}$ occurring in the formulae below are zero.

Proposition B.4 *The numbers $\Omega_{jk\ell}$ can be computed successively as follows.*

(i) *The numbers $\Omega_{1k\ell}$ are all zero, except the following:*

$$\Omega_{1kk} = \frac{u_{k+1} - u_{k-1}}{3}; \quad k = 1, \ldots, m;$$

$$\Omega_{1,k,k+1} = \Omega_{k,k+1,i} = \frac{u_{k+1} - u_k}{6}; \quad k = 1, \ldots, m-1.$$

(ii) *The numbers $\Omega_{2k\ell}$ are all zero, except the following:*

$$\Omega_{2kk} = a_{2,k-1}^2 \Omega_{1,k-1,k-1} - 2a_{2,k-1}a_{2k}\Omega_{1,k-1,k} + a_{2k}^2 \Omega_{1kk};$$
$$k = 1, \ldots, m+1;$$

$$\Omega_{2,k,k+1} = \Omega_{2,k+1,k} = a_{2,k-1}a_{2k}\Omega_{1,k-1,k} - a_{2k}^2 \Omega_{1kk} + a_{2k}a_{2,k+1}\Omega_{1,k,k+1}; \quad i = 1, \ldots, m;$$

$$\Omega_{2,k,k+2} = \Omega_{2,k+2,k} = -a_{2k}a_{2,k+1}\Omega_{1,k,k+1}; \quad k = 1, \ldots, m-1.$$

(iii) *The numbers $\Omega_{3k\ell}$ are all zero, except the following:*

$$\Omega_{3kk} = a_{3,k-1}^2 \Omega_{2,k-1,k-1} - 2a_{3,k-1}a_{3k}\Omega_{2,k-1,k} + a_{3k}^2 \Omega_{2kk};$$
$$k = 1, \ldots, m+2;$$

$$\Omega_{3,k,k+1} = \Omega_{3,k+1,k} = a_{3,k-1}a_{3k}\Omega_{2,k-1,k} - a_{3,k-1}a_{3,k+1}\Omega_{2,k-1,k+1} - a_{3k}^2 \Omega_{2kk} + a_{3k}a_{3,k+1}\Omega_{2,k,k+1}; \quad k = 1, \ldots, m+1;$$

$$\Omega_{3,k,k+2} = \Omega_{3,k+2,k} = a_{3,k-1}a_{3,k+1}\Omega_{2,k-1,k+1} - a_{3k}a_{3,k+1}\Omega_{2,k,k+1} + a_{3k}a_{3,k+2}\Omega_{2,k,k+2}; \quad k = 1, \ldots, m;$$

$$\Omega_{3,k,k+3} = \Omega_{3,k+3,k} = -a_{3k}a_{3,k+2}\Omega_{2,k,k+2}; \quad k = 1, \ldots, m-1.$$

Proof As the computations in the various cases are similar, we do some of them, leaving the verification of the rest to the reader.

Using the basic recursion formula, we have

$$\begin{aligned}
\Omega_{1kk} &= \int B_{1k}(x)B_{1k}(x)\,dx \\
&= \int \left(\frac{x - u_{k-1}}{u_k - u_{k-1}} B_{0,k-1}(x) + \frac{u_{k+1} - x}{u_{k+1} - u_k} B_{0,k}(x) \right)^2 dx \\
&= \big(\text{using that } B_{0,k-1}(x)B_{0,k}(x) = 0 \text{ and } (B_{0,k}(x))^2 = B_{0,k}(x)\big)
\end{aligned}$$

$$= \int \left[\left(\frac{x - u_{k-1}}{u_k - u_{k-1}} \right)^2 B_{0,k-1}(x) + \left(\frac{u_{k+1} - x}{u_{k+1} - u_k} \right)^2 B_{0,k}(x) \right] dx$$

$$= \int_{u_{k-1}}^{u_k} \left(\frac{x - u_{k-1}}{u_k - u_{k-1}} \right)^2 dx + \int_{u_k}^{u_{k+1}} \left(\frac{u_{k+1} - x}{u_{k+1} - u_k} \right)^2 dx$$

$$= \frac{u_k - u_{k-1}}{3} + \frac{u_{k+1} - u_k}{3} = \frac{u_{k+1} - u_{k-1}}{3}.$$

Using Proposition B.1, we get

$$\begin{aligned} \Omega_{2,k,k+1} &= \int B'_{2k}(x) B'_{2,k+1}(x)\, dx \\ &= \int \big(a_{2,k-1} B_{1,k-1}(x) - a_{2k} B_{1k}(x)\big) \\ &\quad \times \big(a_{2k} B_{1k}(x) - a_{2,k+1} B_{1,k+1}(x)\big)\, dx \\ &= (\text{using that } B_{1j}(x) B_{1k}(x) = 0 \text{ unless } k = j-1, j, j+1) \\ &= a_{2,k-1} a_{2k} \Omega_{1,k-1,k} - a_{2k}^2 \Omega_{1kk} + a_{2k} a_{2,k+1} \Omega_{1,k,k+1}. \end{aligned}$$

Using Proposition B.2, we have

$$\begin{aligned} \Omega_{3,k,k+2} &= \int B''_{3k}(x) B''_{3,k+2}(x)\, dx \\ &= \int \big(a_{3,k-1} B'_{2,k-1}(x) - a_{3k} B'_{2k}(x)\big) \\ &\quad \times \big(a_{3,k+1} B'_{2,k+1}(x) - a_{3,k+2} B'_{2,k+2}(x)\big)\, dx \\ &= \big(\text{using that } B'_{2j}(x) B'_{2k}(x) = 0 \text{ unless } k = j-2, j-1, j, j+1, j+2\big) \\ &= a_{3,k-1} a_{3,k+1} \Omega_{2,k-1,k+1} - a_{3k} a_{3,k+1} \Omega_{2,k,k+1} + a_{3k} a_{3,k+2} \Omega_{2,k,k+2}. \end{aligned}$$ □

B.3 Thin Plate Splines

We want to find an $f \in \mathcal{F}$, with $\mathcal{F}$ as defined in Sect. 5.7.1, minimizing the penalized deviance

$$\Delta(f) = D(f) + \lambda \iint \left(\left(\frac{\partial^2 f}{\partial x_1^2} \right)^2 + 2 \left(\frac{\partial^2 f}{\partial x_1 \partial x_2} \right)^2 + \left(\frac{\partial^2 f}{\partial x_2^2} \right)^2 \right) dx_1\, dx_2.$$

The first step is to find the analogue of Lemma B.1.

Lemma B.6 *Let $\mathcal{S}$ be a subset of $\mathcal{F}$, such that*

(i) *Given any real numbers $y_1, \ldots, y_m$, there is a function $s \in \mathcal{S}$, such that $s(z_{1k}, z_{2k}) = y_k$; $k = 1, \ldots, m$;*
(ii) *For any $s \in \mathcal{S}$ and any $h \in \mathcal{F}$ with $h(z_{1k}, z_{2k}) = 0$; $k = 1, \ldots, m$,*

$$\iint \left(\frac{\partial^2 s}{\partial x_1^2} \frac{\partial^2 h}{\partial x_1^2} + 2 \frac{\partial^2 s}{\partial x_1 \partial x_2} \frac{\partial^2 h}{\partial x_1 \partial x_2} + \frac{\partial^2 s}{\partial x_2^2} \frac{\partial^2 h}{\partial x_2^2} \right) dx_1\, dx_2 = 0. \tag{B.25}$$

Then for any $f \in \mathcal{F}$ there is an $s \in \mathcal{S}$, such that $\Delta(s) \le \Delta(f)$.

Proof Entirely similar to the proof of Lemma B.1. □

The next thing we did in the one-dimensional case was to carry out two integration by parts in the counterpart to (B.25), which led us to the condition

$$\frac{d^4}{dx^4} s(x) = 0, \quad \text{for } x \neq z_1, \ldots, z_m. \tag{B.26}$$

Partial integration is a bit more involved in R^2 and we leave the calculations until the proof of Lemma B.7. There it is shown that the analogue of (B.26) is

$$\frac{\partial^4 s}{\partial x^4} + 2 \frac{\partial^4 s}{\partial x^2 \partial y^2} + \frac{\partial^4 s}{\partial y^4} = 0, \quad \text{for } (x_1, x_2) \neq (z_{11}, z_{21}), \ldots, (z_{1m}, z_{2m}). \tag{B.27}$$

However, it turns out that in order to handle the technical details of the integration by parts, things become much easier if we have an explicit expression for s. Therefore, we interchange the order in comparison with Lemmas B.2 and B.3. Thus we first figure out what $s(x_1, x_2)$ ought to look like, based on (B.27) and then prove the important analogue of (B.2).

In the one-dimensional case we found that the functions minimizing the penalized deviance look like

$$s(x) = \frac{1}{12} \sum_{k=1}^{m} d_k |x - z_k|^3 + a_0 + a_1 x. \tag{B.28}$$

They consist of a linear part and a part which is a linear combination of m functions, each of which only depends on the distance from one of the observation points. In R^2, the distance from the point (z_{1k}, z_{2k}) is given by $r_k(x_1, x_2) = \sqrt{(x_1 - z_{1k})^2 + (x_2 - z_{2k})^2}$. A conjecture worth testing is that the minimizing functions in the two-dimensional case look like

$$s(x_1, x_2) = \sum_{k=1}^{m} \delta_k \phi(r_k(x_1, x_2)) + a_0 + a_1 x_1 + a_2 x_2,$$

for some function $\phi(\cdot)$, and where there should be conditions corresponding to (B.4). In the proof of Lemma B.7, it is shown that the solution to (B.27) really is of this form, with $\phi(r) = r^2 \log r^2$. Furthermore, the integration by parts is carried out in detail, leading to a factor corresponding to the $1/12$ in (B.28), and the analogue of Lemma B.2. Here is the result, which due to Lemma B.6 gives an explicit representation for the minimizing functions of the penalized deviance, once we also show that the interpolation property (i) holds.

Lemma B.7 *Let*

$$s(x_1, x_2) = \frac{1}{16\pi} \sum_{k=1}^{m} d_k r_k^2(x_1, x_2) \log r_k^2(x_1, x_2) + a_0 + a_1 x_1 + a_2 x_2, \tag{B.29}$$

where

$$\sum_{k=1}^{m} d_k = 0, \qquad \sum_{k=1}^{m} d_k z_{1k} = 0, \qquad \sum_{k=1}^{m} d_k z_{2k} = 0. \tag{B.30}$$

Then $s \in \mathcal{F}$, *and for any* $h \in \mathcal{F}$,

$$\iint \left(\frac{\partial^2 s}{\partial x_1^2} \frac{\partial^2 h}{\partial x_1^2} + 2 \frac{\partial^2 s}{\partial x \partial x_2} \frac{\partial^2 h}{\partial x_1 \partial x_2} + \frac{\partial^2 s}{\partial x_2^2} \frac{\partial^2 h}{\partial x_2^2} \right) dx_1\, dx_2 = \sum_{k=1}^{m} d_k h(z_{1k}, z_{2k}). \tag{B.31}$$

Proof This proof is quite long and some knowledge of calculus of several variables is required. However, only basic results are used.

The starting point is the expression (B.25) in Proposition B.6:

$$\iint \left(\frac{\partial^2 s}{\partial x_1^2} \frac{\partial^2 h}{\partial x_1^2} + 2 \frac{\partial^2 s}{\partial x_1 \partial x_2} \frac{\partial^2 h}{\partial x_1 \partial x_2} + \frac{\partial^2 s}{\partial x_2^2} \frac{\partial^2 h}{\partial x_2^2} \right) dx_1\, dx_2 = 0. \tag{B.32}$$

In the one-dimensional case we carried out two successive integrations by parts and we shall do this here too. Since not every reader may be up to date on integration by parts in two dimensions, let us recall the necessary results. First we need the concept of integrating along a curve in R^2. Such a curve C may be represented as a mapping $t \mapsto (x_1(t), x_2(t))$, where the parameter t belongs to some interval in R and as t traverses this interval the point $(x_1(t), x_2(t))$ moves along C. The analogy of the fundamental relation between differentiation and integration in one dimension,

$$\int_a^b f'(x)\, dx = f(b) - f(a),$$

is Gauss' theorem,

$$\iint_A \left[\frac{\partial f}{\partial x_1}(x_1, x_2) + \frac{\partial g}{\partial x_2}(x_1, x_2) \right] dx_1\, dx_2 = \int_C [f(x_1, x_2)\, dx_2 - g(x_1, x_2)\, dx_1].$$

Here A is a region in R^2 enclosed by a curve C. The right hand side is a shorthand notation for

$$\int_\alpha^\beta [f(x_1(t), x_2(t))x_2'(t) - g(x_1(t), x_2(t))x_1'(t)]\,dt,$$

where the curve C is represented as $t \mapsto (x_1(t), x_2(t))$, $\alpha \le t \le \beta$, and the orientation of the curve is such that the region A lies on the left side of the curve as t moves from α to β (since the curve is closed, we have that $x_1(\alpha) = x_1(\beta)$, $x_2(\alpha) = x_2(\beta)$). It is not required that $x_1(t)$ and $x_2(t)$ are continuously differentiable, only that the set of points where $x_1'(t)$ and $x_2'(t)$ fail to exist is finite. The curve C can be quite involved. We shall need to consider the situation where A is a large disc in R^2 and excluded from A are m smaller discs centered around the points $(z_{11}, z_{21}), \ldots, (z_{1m}, z_{2m})$ inside the large disc. We can then construct a curve C moving along the circle which is the border of the large disc, then at certain points following a path to one of the discs around a point (z_{1k}, z_{2k}), move along the circle around the point and then return to the large circle along the same path. Finally, after visiting each of the interior discs, the curve ends where it started on the outer circle. The integrals over paths to and from the outer circle to one of the inner circles will cancel out, since the paths are traversed once in each direction and we will get

$$\begin{aligned}\iint_A &\left[\frac{\partial f}{\partial x_1}(x_1, x_2) + \frac{\partial g}{\partial x_2}(x_1, x_2)\right] dx_1\,dx_2 \\ &= \sum_{k=0}^{m} \int_{C_k} [f(x_1, x_2)\,dx_2 - g(x_1, x_2)\,dx_1],\end{aligned} \tag{B.33}$$

where C_0 denotes the large outer circle and $C_1, \ldots, C_m$ the smaller inner circles, remembering that the orientation of the curves are such that the region A lies to the left. If the orientation is the other way round, the integral changes sign.

Using the rule for differentiating a product of two functions, Gauss' theorem gives the formula for integration by parts:

$$\begin{aligned}\iint_A \left(f\frac{\partial u}{\partial x_1} + g\frac{\partial v}{\partial x_2}\right) dx_1\,dx_2 &= \int_C [f u\,dx_2 - g v\,dx_1] \\ &\quad - \iint_A \left(\frac{\partial f}{\partial x_1}u + \frac{\partial g}{\partial x_2}v\right) dx_1\,dx_2.\end{aligned}$$

Here f, g, u and v are four continuously differentiable functions on A.

Let us now consider the integral in (B.32). The integration area is potentially the whole of R^2, but the integral is to be understood as a limit. We therefore start by considering the region A consisting of a disc of radius r_0 centered at the origin, large enough to include the points $(z_{11}, z_{21}), \ldots, (z_{1m}, z_{2m})$, from which we exclude the m discs of radius r around each of the points. The integral in (B.32) is to be understood as a limiting value when $r_0 \to \infty$ and $r \to 0$. Letting C denote the curve described above connecting all of the discs, the integration by parts formula gives

$$\iint_A \left(\frac{\partial^2 s}{\partial x_1^2} \frac{\partial^2 h}{\partial x_1^2} + 2 \frac{\partial^2 s}{\partial x_1 \partial x_2} \frac{\partial^2 h}{\partial x_1 \partial x_2} + \frac{\partial^2 s}{\partial x_2^2} \frac{\partial^2 h}{\partial x_2^2} \right) dx_1\, dx_2$$

$$= \iint_A \left(\frac{\partial^2 s}{\partial x_1^2} \frac{\partial^2 h}{\partial x_1^2} + \frac{\partial^2 s}{\partial x_1 \partial x_2} \frac{\partial^2 h}{\partial x_1 \partial x_2} + \frac{\partial^2 s}{\partial x_2 \partial x_1} \frac{\partial^2 h}{\partial x_2 \partial x_1} + \frac{\partial^2 s}{\partial x_2^2} \frac{\partial^2 h}{\partial x_2^2} \right) dx_1\, dx_2$$

$$= \int_C \left[\left(\frac{\partial^2 s}{\partial x_1^2} \frac{\partial h}{\partial x_1} + \frac{\partial^2 s}{\partial x_2 \partial x_1} \frac{\partial h}{\partial x_2} \right) dx_2 - \left(\frac{\partial^2 s}{\partial x_1 \partial x_2} \frac{\partial h}{\partial x_1} + \frac{\partial^2 s}{\partial x_2^2} \frac{\partial h}{\partial x_2} \right) dx_1 \right]$$

$$- \iint_A \left(\frac{\partial^3 s}{\partial x_1^3} \frac{\partial h}{\partial x_1} + \frac{\partial^3 s}{\partial x_2 \partial x_1^2} \frac{\partial h}{\partial x_2} + \frac{\partial^3 s}{\partial x_1 \partial x_2^2} \frac{\partial h}{\partial x_1} + \frac{\partial^3 s}{\partial x_2^3} \frac{\partial h}{\partial x_2} \right) dx_1\, dx_2.$$

Integrating by parts once more, we have

$$\iint_A \left(\frac{\partial^3 s}{\partial x_1^3} \frac{\partial h}{\partial x_1} + \frac{\partial^3 s}{\partial x_2 \partial x_1^2} \frac{\partial h}{\partial x_2} + \frac{\partial^3 s}{\partial x_1 \partial x_2^2} \frac{\partial h}{\partial x_1} + \frac{\partial^3 s}{\partial x_2^3} \frac{\partial h}{\partial x_2} \right) dx_1\, dx_2$$

$$= \int_C \left[\left(\frac{\partial^3 s}{\partial x_1^3} h + \frac{\partial^3 s}{\partial x_1 \partial x_2^2} h \right) dx_2 - \left(\frac{\partial^3 s}{\partial x_1^2 \partial x_2} h + \frac{\partial^3 s}{\partial x_2^3} h \right) dx_1 \right]$$

$$- \iint_A \left(\frac{\partial^4 s}{\partial x_1^4} + 2 \frac{\partial^4 s}{\partial x_1^2 \partial x_2^2} + \frac{\partial^4 s}{\partial x_2^4} \right) h\, dx_1\, dx_2.$$

As in the one-dimensional case, we want the last integral to be equal to zero. This will happen if the function $s(x_1, x_2)$ satisfies

$$\frac{\partial^4 s}{\partial x_1^4} + 2 \frac{\partial^4 s}{\partial x_1^2 \partial x_2^2} + \frac{\partial^4 s}{\partial x_2^4} = 0, \tag{B.34}$$

for $(x_1, y_1) \neq (z_{11}, z_{21}), \ldots, (z_{1m}, z_{2m})$. Recalling the Laplace operator,

$$\Delta f = \frac{\partial^2 f}{\partial x_1^2} + \frac{\partial^2 f}{\partial x_2^2},$$

we observe that (B.34) may be written as

$$\Delta(\Delta s) = 0. \tag{B.35}$$

At the same point in the one-dimensional case, we concluded that the solution to the corresponding equation was a piecewise cubic polynomial. However, it is not immediately clear what a function satisfying (B.35) looks like. It seems natural to begin by putting $S(x_1, x_2) = \Delta s(x_1, x_2)$ and try to solve the equation $\Delta S = 0$. If this succeeds, we could find s by solving $\Delta s = S$. We note that $\Delta S = 0$ is the well-known Laplace's equation. This equation does not permit explicit solutions in general, only in special cases, for instance when $S(x_1, x_2)$ depends only on the distance from the origin. Now, Δs should correspond to $s''(x)$ in the one-dimensional case and $s''(x)$ has a particularly simple form, being a linear combination of the distances from the points $z_1, \ldots, z_m$. We may try something similar in two dimensions by assuming that $S(x_1, x_2) = \sum_{k=1}^{m} \delta_k \phi(r_k(x_1, x_2))$, where $r_k(x_1, x_2) = \sqrt{(x_1 - z_{1k})^2 + (x_2 - z_{2k})^2}$. To simplify the calculations slightly, we may as well write $\psi(q_k(x_1, x_2))$ instead of $\phi(r_k(x_1, x_2))$, where $q_k(x_1, x_2) = r_k^2(x_1, x_2)$. We then have

$$\begin{aligned}
\Delta S(x_1, x_2) &= \sum_{k=1}^{m} \delta_k \left(\psi''(q_k(x_1, x_2)) \left(\frac{\partial q_k}{\partial x} \right)^2 + \psi'(q_k(x_1, x_2)) \frac{\partial^2 q_k}{dx_1^2} \right. \\
&\quad \left. + \psi''(q_k(x_1, x_2)) \left(\frac{\partial q_k}{\partial x_2} \right)^2 + \psi'(q_k(x_1, x_2)) \frac{\partial^2 q_k}{dx_2^2} \right) \\
&= \sum_{k=1}^{m} \delta_k (4\psi''(q_k(x_1, x_2))(x_1 - z_{1k})^2 + 2\psi'(q_k(x_1, x_2)) \\
&\quad + 4\psi''(q_k(x_1, x_2))(x_2 - z_{2k})^2 + 2\psi'(q_k(x_1, x_2))) \\
&= \sum_{k=1}^{m} \delta_k (4\psi''(q_k(x_1, x_2)) q_k(x_1, x_2) + 4\psi'(q_k(x_1, x_2))). \qquad \text{(B.36)}
\end{aligned}$$

Thus we see that if the function $q \mapsto \psi(q)$ satisfies

$$\psi''(q)q + \psi'(q) = 0, \qquad \text{(B.37)}$$

then $\Delta S = 0$. But the left hand side of (B.37) is the derivative of $\psi'(q)q$, and so $\psi'(q)q = a$ for some constant a. But $\psi'(q) = a/q$ gives $\psi(q) = a \log q + b$, for some constant b. Thus we have

$$S(x_1, x_2) = \sum_{k=1}^{m} \delta_k (a + b \log q_k(x_1, x_2)).$$

Possibly we will have conditions corresponding to (B.4), and in that case we may choose a to be anything. Also b could be incorporated into the δ_k. But let us wait with the constants and continue to see what form $s(x_1, x_2)$ may have. Again making a conjecture based on the one-dimensional case, our guess would be

$$s(x_1, x_2) = \sum_{k=1}^{m} \delta_k \varphi'(q_k(x_1, x_2)) + a_0 + a_1 x_1 + a_2 x_2.$$

As in (B.36) we then get

$$\Delta s(x_1,x_2) = \sum_{k=1}^{m} \delta_k (4\varphi''(q_k(x_1,x_2))q_k(x_1,x_2) + 4\varphi'(q_k(x_1,x_2))).$$

Thus if $q \mapsto \varphi(q)$ satisfies

$$\varphi''(q)q + \varphi'(q) = a + b\log q, \tag{B.38}$$

that is

$$\frac{d}{dq}\{q\varphi'(q)\} = a + b\log q, \tag{B.39}$$

for some constants a and b, we will get $\Delta(\Delta s) = 0$. Now the function $q \mapsto q\log q$ has derivative $1 \cdot \log q + q \cdot (1/q) = \log q + 1$. Therefore, if we take $\varphi'(q) = \log q$, then (B.39) is satisfied with $a = b = 1$. But that makes us realize that if we take $\varphi(q) = q\log q$, then $\varphi'(q) = \log q + 1$ and so (B.39) is satisfied with $a = 2$ and $b = 1$. Thus we have found that if we define

$$s(x_1,x_2) = \sum_{k=1}^{m} \delta_k q_k(x_1,x_2)\log(q_k(x_1,x_2)) + a_0 + a_1x_1 + a_2x_2,$$

where $q_k(x_1,x_2) = (x_1 - z_{1k})^2 + (x_2 - z_{2k})^2$, then $s(x_1,x_2)$ will satisfy (B.34). There should be some normalizing constant corresponding to the $1/12$ in front of the sum in the one-dimensional case, but it will turn up below.

We now have to deal with the rest of the terms resulting from the integration by parts. Let us begin with the term

$$\begin{aligned}
&\int_C \left[\left(\frac{\partial^2 s}{\partial x_1^2}\frac{\partial h}{\partial x_1} + \frac{\partial^2 s}{\partial x_2 \partial x_1}\frac{\partial h}{\partial x_2}\right)dx_2 - \left(\frac{\partial^2 s}{\partial x_1 \partial x_2}\frac{\partial h}{\partial x_1} + \frac{\partial^2 s}{\partial x_2^2}\frac{\partial h}{\partial x_2}\right)dx_1\right] \\
&= \int_C \left[\left(\sum_{k=1}^{m} \delta_k \left(2\log q_k(x_1,x_2) + 4\frac{(x_1 - z_{1k})^2}{q_k(x_1,x_2)}\right)\frac{\partial h}{\partial x_1}\right.\right. \\
&\quad \left. + \sum_{k=1}^{m} \delta_k 4\frac{(x_1 - z_{1k})(x_2 - z_{2k})}{q_k(x_1,x_2)}\frac{\partial h}{\partial x_2}\right)dx_2 \\
&\quad - \left(\sum_{k=1}^{m} \delta_k 4\frac{(x_1 - z_{1k})(x_2 - z_{2k})}{q_k(x_1,x_2)}\frac{\partial h}{\partial x_1}\right. \\
&\quad \left.\left. + \sum_{k=1}^{m} \delta_k \left(2\log q_k(x_1,x_2) + 4\frac{(x_2 - z_{2k})^2}{q_k(x_1,x_2)}\right)\frac{\partial h}{\partial x_2}\right)dx_1\right].
\end{aligned} \tag{B.40}$$

We let $r_0 > 0$ and $r > 0$ and introduce the curves

$$\begin{aligned}
C_0 &= \{(x_1,x_2) \in R^2 : x_1^2 + x_2^2 = r_0\}, \\
C_k &= \{(x_1,x_2) \in R^2 : (x_1 - z_{1k})^2 + (x_2 - z_{2k})^2 = r\}; \quad k = 1,\ldots,m.
\end{aligned}$$

Here r_0 is assumed to be large enough for the circles C_k, $k = 1, \ldots, m$ to lie within the circle C_0. As explained above, we can then divide the integral (B.40) as

$$\iint_C = \iint_{C_0} + \iint_{C_1} + \cdots + \iint_{C_m}.$$

The outer circle C_0 is represented by

$$x_1(t) = r_0 \cos t, \qquad x_2(t) = r_0 \sin t, \quad 0 \le t \le 2\pi.$$

We get that

$$\begin{aligned} q_k(x_1(t), x_2(t)) &= (r_0 \cos t - z_{1k})^2 + (r_0 \sin t - z_{2k})^2 \\ &= r_0^2 - 2r_0 z_{1k} \cos t - 2r_0 z_{2k} \sin t + z_{1k}^2 + z_{2k}^2. \end{aligned}$$

Using the well-known expansions

$$\begin{aligned} \log(1-x) &= -x + O(x^2), \quad |x| < 1, \\ \frac{1}{1-x} &= 1 + x + O(x^2), \quad |x| < 1, \end{aligned}$$

we get

$$\begin{aligned} \log q_k(x_1(t), x_2(t)) &= \log r_0^2 + \log\left(1 - 2\frac{z_{1k}}{r_0}\cos t - 2\frac{z_{2k}}{r_0}\sin t + \frac{z_{1k}^2 + z_{2k}^2}{r_0^2}\right) \\ &= \log r_0^2 - 2\frac{z_{1k}}{r_0}\cos t - 2\frac{z_{2k}}{r_0}\sin t + O(1/r_0^2) \end{aligned}$$

and

$$\begin{aligned} \frac{(x_1(t) - z_{1k})^2}{q_k(x_1(t), x_2(t))} &= \frac{\cos^2 t - 2\frac{z_{1k}}{r_0}\cos t + \frac{z_{1k}^2}{r_0^2}}{1 - 2\frac{z_{1k}}{r_0}\cos t - 2\frac{z_{2k}}{r_0}\sin t + \frac{z_{1k}^2 + z_{2k}^2}{r_0^2}} \\ &= \cos^2 t - 2\frac{z_{1k}}{r_0}\cos t + 2\frac{z_{1k}}{r_0}\cos^3 t + O(1/r_0^2). \end{aligned}$$

Thus, if we assume

$$\sum_{k=1}^{m} \delta_k = 0, \qquad \sum_{k=1}^{m} \delta_k z_{1k} = 0, \qquad \sum_{k=1}^{m} \delta_k z_{2k} = 0, \tag{B.41}$$

we get

$$\sum_{k=1}^{m} \delta_k \left(2 \log q_k(x_1(t), x_2(t)) + 4\frac{(x_1(t) - z_{1k})^2}{q_k(x_1(t), x_2(t))}\right)$$

$$
\begin{aligned}
&= \sum_{k=1}^{m} \delta_k \left(2\left(\log r_0^2 - 2\frac{z_{1k}}{r_0}\cos t - 2\frac{z_{2k}}{r_0}\sin t \right) \right. \\
&\quad \left. + 4\left(\cos^2 t - 2\frac{z_{1k}}{r_0}\cos t + 2\frac{z_{1k}}{r_0}\cos^3 t + O(1/r_0^2) \right) \right) \\
&= O(1/r_0^2).
\end{aligned}
$$

Likewise,

$$
\sum_{k=1}^{m} \delta_k 4 \frac{(x_1(t) - z_{1k})(x_2(t) - z_{2k})}{q_k(x_1(t), x_2(t))} = O(1/r_0^2),
$$

$$
\sum_{k=1}^{m} \delta_k \left(2\log q_k(x_1(t), x_2(t)) + 4\frac{(x_2(t) - z_{2k})^2}{q_k(x_1(t), x_2(t))} \right) = O(1/r_0^2).
$$

Since by assumption $\partial h/\partial x_1$ and $\partial h/\partial x_2$ are $O(1)$, we get

$$
\begin{aligned}
\int_{C_0} \cdots &= \int_0^{2\pi} [(O(1/r_0^2)O(1) + O(1/r_0^2)O(1))x_2'(t) \\
&\quad - (O(1/r_0^2)O(1) + O(1/r_0^2)O(1))x_1'(t)]\, dt \\
&= \int_0^{2\pi} O(1/r_0^2) r_0 \, dt = O(1/r_0).
\end{aligned}
$$

Thus, due to the conditions (B.41), the integral over C_0 tends to zero as $r_0 \to \infty$.

Consider now one of the smaller circles C_j, represented by

$$
x_1(t) = z_{1j} + r\cos t, \qquad x_2(t) = z_{2j} + r\sin t, \quad 0 \le t \le 2\pi.
$$

Keeping in mind that the orientation of C_j is opposite to C_0, the integral is

$$
\begin{aligned}
&\sum_{k=1}^{m} \delta_k \int_{C_j} \left(4\frac{(x_1 - z_{1k})(x_2 - z_{2k})}{q_k(x_1, x_2)} \frac{\partial h}{\partial x_1} \right. \\
&\quad \left. + \left(2\log q_k(x_1, x_2) + 4\frac{(x_2 - z_{2k})^2}{q_k(x_1, x_2)} \right) \frac{\partial h}{\partial x_2} \right) dx_1 \\
&\quad - \left(\left(2\log q_k(x_1, x_2) + 4\frac{(x_1 - z_{1k})^2}{q_k(x_1, x_2)} \right) \frac{\partial h}{\partial x_1} \right. \\
&\quad \left. + 4\frac{(x_1 - z_{1k})(x_2 - z_{2k})}{q_k(x_1, x_2)} \frac{\partial h}{\partial x_2} \right) dx_2.
\end{aligned}
$$

For the terms in the sum where $k \neq j$, due to the continuity of the integrand, the integral becomes

$$\int_0^{2\pi} O(1) r\, dt = O(r),$$

and thus tends to zero as $r \to 0$. For $k = j$ the integral is

$$\int_0^{2\pi} [(2\log r^2 + \sin^2 t) O(1) + 4 \sin t \cos t\, O(1) + 4 \sin t \cos t\, O(1)$$
$$+ (2\log r^2 + \cos^2 t) O(1)] r\, dt = O(r \log r),$$

which also tends to zero as $r \to 0$.

We now turn to the final piece left over from the repeated integration by parts:

$$\int \left[\left(\frac{\partial^3 s}{\partial x_1^3} + \frac{\partial^3 s}{\partial x_1 \partial x_2^2} \right) h\, dx_2 - \left(\frac{\partial^3 s}{\partial x_1^2 \partial x_2} + \frac{\partial^3 s}{\partial x_2^3} \right) h\, dx_1 \right]$$
$$= \int_C \left[\frac{\partial}{\partial x_1} \Delta s h\, dx_2 - \frac{\partial}{\partial x_2} \Delta s h\, dx_1 \right]$$
$$= \int_C \left[8 \sum_{k=1}^{m} \delta_k \frac{x_1 - z_{1k}}{q_k(x_1, x_2)} h\, dx_2 - 8 \sum_{k=1}^{m} \delta_k \frac{x_2 - z_{2k}}{q_k(x_1, x_2)} h\, dx_1 \right].$$

As above we split the integral into the integral over the outer circle C_0 and the inner circles C_k. Beginning with C_0 we get in the same manner as before

$$\frac{x_1(t) - z_{1k}}{q_k(x_1(t), x_2(t))}$$
$$= \frac{1}{r_0^2} \left(1 - 2\frac{z_{1k}}{r_0} \cos t - 2\frac{z_{2k}}{r_0} \sin t + O(1/r_0^2) \right) (r_0 \cos t - z_{1k})$$
$$= \frac{1}{r_0} \left(\cos t - \frac{z_{1k}}{r_0} - 2\frac{z_{1k}}{r_0} \cos^2 t - 2\frac{z_{2k}}{r_0} \sin t \cos t + O(1/r_0^2) \right).$$

Thus, due to (B.41), we have

$$\sum_{k=1}^{m} \delta_k \frac{x_1(t) - z_{1k}}{q_k(x_1(t), x_2(t))} = O(1/r_0^3).$$

Similarly,

$$\sum_{k=1}^{m} \delta_k \frac{x_2(t) - z_{2k}}{q_k(x_1(t), x_2(t))} = O(1/r_0^3).$$

Since $\partial h/\partial x_1$ and $\partial h/\partial x_2$ are bounded, the mean value theorem of differential calculus implies that $h(x_1(t), x_2(t)) = O(r_0)$. Therefore the integral over C_0 becomes

$$\int_0^{2\pi} O(1/r_0^3)O(r_0)r_0\,dt = O(1/r_0).$$

Thus the integral tends to zero as $r_0 \to \infty$.

Finally, the integral over one of the small circles C_j is

$$8\sum_{k=1}^{m}\delta_k \int_{C_j}\left[\frac{x_2 - z_{2k}}{q_k(x_1,x_2)} h\,dx_1 - \frac{x_1 - z_{1k}}{q_k(x_1,x_2)} h\,dx_2\right].$$

As before, the integral is $O(r)$ when $k \neq j$. When $k = j$, the integral becomes

$$\int_0^{2\pi}\left(\frac{r\sin t}{r^2}(-r\sin t) - \frac{r\cos t}{r^2} r\cos t\right) h(x_1(t), x_2(t))\,dt$$
$$= -\int_0^{2\pi} h(x_1(t), x_2(t))\,dt.$$

Since

$$2\pi \min_{0\le t\le 2\pi} h(x_1(t), x_2(t)) \le \int_0^{2\pi} h(x_1(t), x_2(t))\,dt \le 2\pi \max_{0\le t\le 2\pi} h(x_1(t), x_2(t)),$$

due to the continuity of h, the integral tends to $-2\pi h(z_{1j}, z_{2j})$ as $r \to 0$. Summing everything up from the beginning, we get that the limit of the integral

$$\iint_A\left(\frac{\partial^2 s}{\partial x_1^2}\frac{\partial^2 h}{\partial x_1^2} + 2\frac{\partial^2 s}{\partial x_1\partial x_2}\frac{\partial^2 h}{\partial x_1\partial x_2} + \frac{\partial^2 s}{\partial x_2^2}\frac{\partial^2 h}{\partial x_2^2}\right) dx_1\,dx_2,$$

as $r_0 \to \infty$ and $r \to 0$ is $\sum_{k=0}^{m} 16\pi\delta_k h(z_{1k}, z_{2k})$. Finally, we put $d_k = 16\pi\delta_k$ to get (B.31).

To conclude, note that the only properties of h that we used in the proof was that the first partial derivatives are continuous and bounded. Since s satisfies these requirements, we may take $h = s$ in the proof above and deduce that

$$\iint\left(\left(\frac{\partial^2 s}{\partial x_1^2}\right)^2 + 2\left(\frac{\partial^2 s}{\partial x_1\partial x_2}\right)^2 + \left(\frac{\partial^2 s}{\partial x_2^2}\right)^2\right) dx_1\,dx_2 = \sum_{k=1}^{m} d_k s(z_{1k}, z_{2k}).$$

In particular, $s \in \mathcal{F}$. □

A function of the form (B.29) satisfying (B.30) is called a *thin plate spline*. The name comes from the fact that the integral in the penalty term is an approximation to the strain energy of a deformed infinite thin plate. The integral in the penalty term in one dimension has a similar interpretation in terms of the strain energy when bending a thin rod.

We can now find an explicit expression for the penalty term.

Lemma B.8 *With* $s(x_1, x_2)$ *as in* (B.29),

$$\iint \left(\left(\frac{\partial^2 s}{\partial x_1^2} \right)^2 + 2 \left(\frac{\partial^2 s}{\partial x_1 \partial x_2} \right)^2 + \left(\frac{\partial^2 s}{\partial x_2^2} \right)^2 \right) dx_1 \, dx_2$$

$$= \frac{1}{16\pi} \sum_{j=1}^{m} \sum_{k=1}^{m} d_j d_k r_k^2(z_{1j}, z_{2j}) \log r_k^2(z_{1j}, z_{2j}).$$

Proof Take $h = s$ in (B.31) and use (B.30). □

Finally, we prove the interpolation property (i) in Lemma B.6 for the family $\mathcal{S}$ of thin plate splines with knots $(z_{11}, z_{21}), \ldots, (z_{1m}, z_{2m})$.

Theorem B.4 *Let* $(z_{11}, z_{21}), \ldots, (z_{1m}, z_{2m})$ *be* m *distinct points in* R^2 *which do not lie on a straight line, and let* $y_1, \ldots, y_m$ *be any real numbers. Then there exists a unique thin plate spline* $s(x_1, x_2)$, *such that* $s(z_{1j}, z_{2j}) = y_j$, $j = 1, \ldots, m$.

Proof With minor modifications, the proof can be copied from the proof of Theorem B.1. The elements in the matrix **E** are now instead

$$e_{jk} = r_k^2(z_{1j}, z_{2j}) \log r_k^2(z_{1j}, z_{2j})/(16\pi).$$

Of course, $e_{jk} = e_{kj}$. Instead of the single vector $\mathbf{z}$, we now have two vectors

$$\mathbf{z}_1 = \begin{bmatrix} z_{11} \\ z_{12} \\ \vdots \\ z_{1m} \end{bmatrix}, \qquad \mathbf{z}_2 = \begin{bmatrix} z_{21} \\ z_{22} \\ \vdots \\ z_{2m} \end{bmatrix},$$

and (B.9) becomes

$$\begin{bmatrix} \mathbf{E} & \mathbf{1} & \mathbf{z}_1 & \mathbf{z}_2 \\ \mathbf{1}' & 0 & 0 & 0 \\ \mathbf{z}_1{}' & 0 & 0 & 0 \\ \mathbf{z}_2{}' & 0 & 0 & 0 \end{bmatrix} \begin{bmatrix} \mathbf{d} \\ a_0 \\ a_1 \\ a_2 \end{bmatrix} = \begin{bmatrix} \mathbf{y} \\ 0 \\ 0 \\ 0 \end{bmatrix}.$$

The rest of the proof is entirely similar and ends by concluding that $a_0 = a_1 = a_2 = 0$, since the points $(z_{11}, z_{21}), \ldots, (z_{1m}, z_{2m})$ are non-collinear. □

Appendix C
Some SAS Syntax

There are many software solutions for non-life insurance pricing with GLMs, ranging from specialized software where you more or less just "push the button", to manual programming of the numerics in a low-level language. The examples in this book were generated using the SAS system, which falls somewhere in between these two extremes. We will provide some sample code in this appendix, as an aid for the reader who is already familiar with the basics of the SAS system. For full information on SAS, we refer to the system manuals, see e.g. [SASI08].

We have chosen SAS because it is widely used in the insurance industry, but we would also like to mention the possibility of using the software R—popular in the academic world—or Matlab. For instance, see the book by Wood [Wo06] for a treatment of generalized additive models with R.

C.1 Parameter Estimation with Proc Genmod

Here we give the code for claim frequency estimation that we have used for Example 2.5. We assume that we already have a SAS table called moped containing the data of Table 1.2. First chose a *base cell* by changing the name of its levels to 'BASE', the point being that 'BASE' comes after the others in the alphanumerical ordering that SAS uses for sorting. The level with the value that comes last becomes the base level in SAS. In the example, everything that starts with "mop" is an optional name entered by the user.

```
data moped;
    set moped;
    if mopclass='1' then mopclass='BASE';
    if mopage='2' then mopage='BASE';
    if mopzone='4' then mopzone='BASE';
run;
```

E. Ohlsson, B. Johansson, *Non-Life Insurance Pricing with Generalized Linear Models*, EAA Lecture Notes,
DOI 10.1007/978-3-642-10791-7, © Springer-Verlag Berlin Heidelberg 2010

```
proc genmod data=moped;
    class mopclass mopage mopzone;
    model mopfreq = mopclass mopage mopzone
                  / dist=poisson link=log;
    ods output ParameterEstimates=mopparam;
    weight mopdur;
    title 'Claim frequency with Proc Genmod';
run;

data moprel (keep= parameter level1 rel);
    set mopparam;
    if Parameter='mopclass' and level1='BASE' then level1='1';
    if Parameter='mopage' and level1='BASE' then level1='2';
    if Parameter='mopzone' and level1='BASE' then level1='4';
    if parameter='Intercept' or parameter='Scale' then delete;
    rel = exp(estimate);
    format rel 7.2;
run;
```

The `proc genmod` step is where the GLM analysis is performed. The statement `ods output ParameterEstimates=mopparam` generates a SAS table called `mopparam`, that contains the parameter estimates. With the `weight` statement, the weight is set to the duration. The last data step makes the transition from the β parameters to the multiplicative relativities γ and resets the level names to the original ones.

C.2 Estimation of ϕ and Testing

Proc Genmod uses as default the ML estimator of ϕ when a gamma distribution is chosen. To get the Pearson estimator $\hat{\phi}_X$ in (3.9) we have to add the option `pscale` at the end of the `model` statement.

```
model mopsev = x y z / dist=gamma link=log pscale;
```

Proc Genmod produces effect tests like those in Table 3.1 in its so called "Type 3 analysis". If we add the `type3` option to the model statement, SAS will produce a table like Table C.1, where we have also used the `pscale` option.

Here the `Source` column gives the name of the excluded rating factor and `Num DF` is the number of excluded parameters $f_r - f_s$, i.e., the degrees of freedom in the LRT, while `Den DF` is the degrees of freedom $f_m - f_r$ in $\hat{\phi}$. The F-statistic displayed in the column `F Value` will be explained below. The LRT is given in the `Chi-Square` column. Finally, `Pr>F` and `Pr>ChiSq` give the p-values for these two tests.

Table C.1 Moped insurance: output from SAS Proc Genmod. Test of excluding one variable at a time. Multiplicative gamma model

LR Statistics For Type 3 Analysis

Sourse	Num DF	Den DF	F Value	Pr > F	Chi-Square	Pr > ChiSq
mopclass	1	16	122.71	<.0001	122.71	<.0001
mopage	1	16	79.91	<.0001	79.91	<.0001
mopzone	6	16	1.30	0.3131	7.79	0.2539

The so called F-statistic is obtained by dividing the LRT by its degrees of freedom:

$$F \doteq \frac{(D^*(y, \hat{\mu}^{(s)}) - D^*(y, \hat{\mu}^{(r)}))/(f_r - f_s)}{\hat{\phi}_X}.$$

The F-statistic is familiar from the ordinary linear model, where it is F-distributed. With other GLMs, we suggest that the LRT is used in the first place.

C.3 SAS Syntax for Arbitrary Deviance*

SAS Proc Genmod has built-in options for the most common EDMs (normal, Poisson, gamma, binomial, inverse Gaussian, and more); if, e.g., we want to choose the Poisson variance function we add the option `dist=poisson` to the model statement. Models that are not standard in Proc Genmod can be analyzed by specifying the variance function and the deviance in special "programming statements". We illustrate the procedure by presenting code that gives the same result as if we had chosen the option `dist=poisson`. The intention is that the reader should be able to adjust this code as needed, e.g., in the case of Tweedie models with $1 < p < 2$. Note that `_mean_` and `_resp_` are *automatic variables* provided by the system, representing μ and y, respectively, and do not have to be specified by the user.

```
proc genmod data=moped;
    class mopclass mopage;
         a=_mean_;
         y=_resp_;
         if y=0 then d=2*a; else d=2*(y*log(y/a)-(y-a));
         variance var=a;
         deviance dev=d;
    model mopfreq = mopclass mopage /link=log pscale;
    weight dur;
run;
```

C.4 Backfitting of MLFs

We will not give all SAS syntax for performing the backfitting algorithm in Sect. 4.2.2, since this is mostly a question of straight-forward data steps. However, we shall give a hint on how to easily obtain $\mu_i = \mu\gamma_i$ from the *predicted values* that can be retrieved from Proc Genmod by adding the statement `ods output ObStats=PredVal;`

After an initial Proc Genmod step with this statement included, the μ_i—here called mu—can be computed and added to your data as follows, where `uoff` is the offset variable $\log(u_j)$,

```
data infile;
    merge infile PredVal(keep=pred);
    mu=pred/exp(uoff);
    drop pred uoff;
run;
```

C.5 Fitting GAMs

The SAS procedure Proc GAM fits generalized additive models using smoothing splines and thin plate splines. A drawback is that the procedure, at least at the time of writing, does not provide a `weight` statement, which prohibits a simple analysis of claim frequencies. Hopefully, this will be amended in future versions. For documentation, see [SASI08].

Our own implementation of GAMs was carried out using data step programming and SAS/IML.

C.6 Miscellanea

Interaction between two variables $x1$ and $x2$ may be tested in SAS by including an interaction term in the `model` statement, like this:

```
model y = x1 x2 x1*x2 x3
                    / dist=poisson link=log;
```

Once a significant interaction effect is found we strongly suggest that a new combined rating factor is introduced rather than using the above code.

Offsetting is done in SAS by including the option `offset=z` in the model statement, if z is the offset variable on the linear scale.

Polynomial regression with Proc GenMod requires only some slight changes to the code in Sect. C.1. Assume that we have prepared the same SAS table with the difference that the categorical variable mopage is replaced by a numerical variable called mopyrs, which is the moped age in years: $0, 1, 2, \ldots$. To produce the estimates, representing the effect of moped age using a third degree polynomial, the code is as follows:

```
proc genmod data=moped;
    class mopclass mopzone;
    model mopfreq = mopclass mopzone mopyrs mopyrs*mopyrs
    mopyrs*mopyrs*mopyrs / dist=poisson link=log;
    ods output ParameterEstimates=mopparam;
    weight mopdur;
    title 'Claim frequency with Polynomial regression';
run;
```

Thus in the model statement the mopage variable is replaced by three variables, corresponding to the linear, quadratic and cubic part of the polynomial, and these three variables are not included in the class statement.

In the numerical solving of the ML equations in Proc Genmod, Fisher's method is used in step 1, and Newton-Raphson thereafter, unless one specifies the option Scoring=n in the Model statement, by which Fisher's method is used in the n first steps. It is, of course, futile to try to find the optimal value of n, but if one should encounter problems with the convergence of Proc Genmod, it is worth trying some $n > 1$ in the option Scoring.

References

[An07] Andersen, R.: Modern Methods for Robust Regression. Sage, Thousand Oaks (2007)

[Ba63] Bailey, R.A.: Insurance rates with minimum bias. In: Proceedings of the Casualty Actuarial Society, vol. L (1963)

[BPP84] Beard, R.E., Pentikäinen, T., Pesonen, E.: Risk Theory. Chapman & Hall, London (1984)

[BW92] Brockman, M.J., Wright, T.S.: Statistical motor rating: making efficient use of your data. J. Inst. Actuar. **119**, 457–543 (1992)

[Bü67] Bühlmann, H.: Experience rating and credibility. ASTIN Bull. **4**(3), 336–341 (1967)

[BG05] Bühlmann, H., Gisler, A.: A Course in Credibility Theory and Its Applications. Universitext. Springer, Berlin (2005)

[BS70] Bühlmann, H., Straub, E.: Glaubwürdichkeit für Schadensätze. Bull. Swiss Assoc. Actuar. **70**, 111–133 (1970)

[Ca86] Campbell, M.: An integrated system for estimating the risk premium of individual car models in motor insurance. ASTIN Bull. **16**(2), 165–184 (1986)

[DKG96] Dannenburg, D.R., Kaas, R., Goovaerts, M.J.: Practical Actuarial Credibility Models. Institute of Actuarial Science and Econometrics, Amsterdam (1996)

[DPP94] Daykin, C.D., Pentikäinen, T., Pesonen, M.: Practical Risk Theory for Actuaries. Chapman & Hall, London (1994)

[dBo01] de Boor, C.: A Practical Guide to Splines, revised edn. Springer, Berlin (2001)

[Di05] Dickson, C.M.: Insurance Risk and Ruin. Cambridge University Press, Cambridge (2005)

[GH87] Goovaerts, M.J., Hoogstad, W.J.: Credibility Theory. Survey of Actuarial Studies, vol. 4. Nationale-Nederlanden, Den Haag (1987)

[Gr08] Grgić, V.: Smoothing splines in non-life insurance pricing. Master Thesis, Institute of Mathematical Statistics, Stockholm University (2008)

[GS94] Green, P.J., Silverman, B.W.: Nonparametric Regression and Generalized Linear Models: A Roughness Penalty Approach. Chapman & Hall, London (1994)

[Gu95] Gut, A.: An Intermediate Course in Probability. Springer, Berlin (1995)

[HR96] Haberman, S., Renshaw, A.E.: Generalized linear models and actuarial science. Statistician **45**, 407–436 (1996)

[HT90] Hastie, T.J., Tibshirani, R.J.: Generalized Additive Models. Chapman & Hall, London (1990)

[IJ07] Ismail, N., Jemain, A.A.: Handling overdispersion with negative binomial and generalized Poisson regression models. In: Casualty Actuarial Society Forum, Winter (2007)

[IJ09] Ismail, N., Jemain, A.A.: Comparison of minimum bias and maximum likelihood methods for claim severity. In: Casualty Actuarial Society Forum, Winter (2009)

E. Ohlsson, B. Johansson, *Non-Life Insurance Pricing with Generalized Linear Models*, EAA Lecture Notes,
DOI 10.1007/978-3-642-10791-7, © Springer-Verlag Berlin Heidelberg 2010

[Je74] Jewell, W.S.: Credible means are exact Bayesian for exponential families. ASTIN Bull. **8**, 336–341 (1974)

[Jö87] Jörgensen, B.: Exponential dispersion models. J. R. Stat. Soc. Ser. B **49**, 127–162 (1987)

[Jö97] Jörgensen, B.: The Theory of Dispersion Models. Chapman & Hall, London (1997)

[JS94] Jörgensen, B., Paes de Souza, M.C.: Fitting Tweedie's compound Poisson model to insurance claim data. Scand. Actuar. J. **1994**, 69–93 (1994)

[Ju68] Jung, J.: On automobile insurance rating. ASTIN Bull. **5**(1), 41–48 (1968)

[KPW04] Klugman, S.A., Panjer, H.H., Wilmot, G.E.: Loss Models, from Data to Decisions, 2nd edn. Wiley, New York (2004)

[LN96] Lee, Y., Nelder, J.A.: Hierarchical generalized linear models. J. R. Stat. Soc. Ser. B **58**(4), 619–678 (1996)

[LNP06] Lee, Y., Nelder, J.A., Pawitan, Y.: Generalized Linear Models with Random Effects. Chapman & Hall, London (2006)

[Li93] Lindgren, B.: Statistical Theory, 4th edn. Chapman & Hall, London (1993)

[Ma91] Mack, T.: A simple parametric model for rating automobile insurance of estimating IBNR claims reserves. ASTIN Bull. **21**, 93–109 (1991)

[McC83] McCullagh, P.: Quasi-likelihood functions. Ann. Stat. **11**, 59–67 (1983)

[MN89] McCullagh, P., Nelder, J.A.: Generalized Linear Models, 2nd edn. Chapman & Hall, London (1989)

[Me04] Meng, R.: Estimation of dispersion parameters in GLMs with and without random effects. Institute of Mathematical Statistics, Stockholm University (2004)

[MBL00] Murphy, K.P., Brockman, M.J., Lee, P.K.W.: Using generalized linear models to build dynamic pricing systems for personal lines insurance. In: CAS Winter 2000 Forum (2000)

[NW72] Nelder, J.A., Wedderburn, R.W.M.: Generalized linear models. J. R. Stat. Soc. Ser. A **135**, 370–384 (1972)

[Oh05] Ohlsson, E.: Simplified estimation of structure parameters in hierarchical credibility. In: The 36th ASTIN Colloquium, Zürich (2005). http://www.astin2005.ch/

[Oh08] Ohlsson, E.: Combining generalized linear models and credibility models in practice. Scand. Actuar. J. **2008**, 301–314 (2008)

[OJ06] Ohlsson, E., Johansson, B.: Exact credibility and Tweedie models. ASTIN Bull. **36**(1), 121–133 (2006)

[Re94] Renshaw, A.E.: Modelling the claims process in the presence of covariates. ASTIN Bull. **24**(2), 265–285 (1994)

[Ro91] Robinson, G.K.: That BLUP is a good thing—the estimation of random effects. Stat. Sci. **6**, 15–51 (1991)

[RT97] Rootzén, H., Tajvidi, N.: Extreme value statistics and wind storm losses: a case study. Scand. Actuar. J. **1**, 70–94 (1997)

[Ro02] Ross, S.: A First Course in Probability, 6th edn. Prentice Hall, New York (2002)

[SASI08] SAS Institute: SAS Product Documentation (2008). http://support.sas.com/documentation/index.html

[SS91] Sibson, R., Stone, G.: Computation of thin plate splines. SIAM J. Sci. Stat. Comput. **12**(6), 1304–1313 (1991)

[Su03] Sundberg, R.: Conditional statistical inference and quantification of relevance. J. R. Stat. Soc. Ser. B **65**(1), 299–315 (2003)

[Su80] Sundt, B.: A multi-level hierarchical credibility regression model. Scand. Actuar. J. 25–32 (1980)

[Su87] Sundt, B.: Two credibility regression approaches for the classification of passenger cars in a multiplicative tariff. ASTIN Bull. **17**(1), 42–70 (1987)

[Wo06] Wood, S.: Generalized Additive Models: An Introduction with R. Chapman & Hall, London (2006)

Index

E. Ohlsson, B. Johansson, *Non-Life Insurance Pricing with Generalized Linear Models*, EAA Lecture Notes,
DOI 10.1007/978-3-642-10791-7,